Advances in Heat Transfer Enhancement

Sujoy Kumar Saha • Manvendra Tiwari
Bengt Sundén • Zan Wu

Advances in Heat Transfer Enhancement

Sujoy Kumar Saha
IIEST
Shibpur, Howrah, India

Manvendra Tiwari
Mechanical Engineering Department
Indian Institute of Engineering Science
Howrah, West Bengal, India

Bengt Sundén
Department of Energy Sciences
Lund University
Lund SE, Sweden

Zan Wu
Department of Energy Sciences
Lund University
Lund SE, Sweden

ISBN 978-3-319-29478-0 ISBN 978-3-319-29480-3 (eBook)
DOI 10.1007/978-3-319-29480-3

Library of Congress Control Number: 2016933105

Printed on acid-free paper

This Springer imprint is published by Springer Nature
The registered company is Springer International Publishing AG Switzerland

Contents

Part I Single Phase Flow

1 Introduction ... 3
- 1.1 Heat Exchanger Fundamentals ... 3
- 1.2 Impact of Boundary Layer Thickness on Thermal Resistance ... 5
- 1.3 Influence of Types of Flow on Heat Transfer ... 6
- 1.4 Advantages and Disadvantages of Laminar Flow ... 6

2 Meaning of Improved Heat Exchanger ... 7
- 2.1 Derivation of Basic Equations for Performance Evaluation and Performance Plot ... 8
- 2.2 Plot Interpretation ... 12
- 2.3 How to Use the Proposed Performance Plot ... 13

3 Advances in Passive Techniques ... 15
- 3.1 Effort to Improve Existing Heat Exchangers ... 15
  - 3.1.1 Investigation of Heat Transfer Enhancement by Perforated Helical Twisted-Tapes ... 16
  - 3.1.2 V-Cut Twisted Tape Insert ... 21
- 3.2 Effect of Internally Grooved Shape on Heat Transfer Augmentation ... 24
- 3.3 Twisted Elliptical Tubes ... 26
  - 3.3.1 Heat Transfer Performance ... 26
  - 3.3.2 Frictional Loss Aspect of Twisted Tubes ... 28

4 Advances in Compound Techniques (Fourth Generation Heat Transfer Technology) ... 31
- 4.1 Helical-Ribbed Tube with Double Twisted Tape Inserts ... 31
  - 4.1.1 Heat Transfer Performance of Ribbed Tube with Twin Twisted Tape ... 32
  - 4.1.2 Effect on Friction Factor ... 32
  - 4.1.3 Performance Evaluation of Ribbed Tube with Twin Twisted Tape ... 33

4.1.4 Empirical Correlations ... 35
4.1.5 Influence of Combined Non-uniform Wire Coil and Twisted Tape Inserts ... 35
4.1.6 Heat Transfer Augmentation and Comparison ... 36
4.1.7 Dimpled Tube with Twisted Tape Inserts ... 37
4.1.8 Influence of Twist Ratio on Heat Transfer and Frictional Losses ... 38
4.1.9 Integral Type Wall Roughness with Wavy Strip Inserts ... 41

5 Nano-Fluids, Next-Generation Heat Transfer ... 45
5.1 Classification of Nano-Fluids System ... 45
5.2 Distinct Features of Nano-Fluids ... 47
5.3 Preparation Methods for Nano-Fluids ... 47
5.4 Application of Nano-Fluids in Automobile Heat Exchangers as Coolant ... 48
5.4.1 Effect of Augmentation of Nano-Particle Concentration on Radiator Cooling Performance ... 50
5.4.2 Effect of Fluid Inlet Temperature on Heat Transfer Performance of the Automobile Radiator ... 51

6 Effect of Ultrasounds on Thermal Exchange ... 53
6.1 Understanding Enhancement Mechanism ... 53

7 Conclusions ... 59
Bibliography ... 60

Part II Enhancement of Heat Transfer in Two-Phase Flow

8 Introduction ... 77

9 Passive Techniques ... 81
9.1 Surface Coatings ... 81
9.1.1 Macro/Microporous Coatings ... 81
9.1.2 Nanoscale Surface Coating ... 83
9.2 Roughened and Finned Surfaces ... 90
9.3 Insert Devices ... 100
9.4 Curved Geometry ... 102
9.5 Additives ... 102

10 Active Techniques ... 111

11 Additional Remarks ... 119

12 Conclusions and Future Work ... 123
Bibliography ... 124

Nomenclature

A	Area
A	Area
A_{ni}	Nominal surface area based on fin root diameter
D	Diameter
d	Diameter
d_i	Fin root diameter
d_o	Outer diameter
E	Electric field strength
e	Fin height
f	Friction factor
$\mathbf{f}_e$	EHD force density
$\boldsymbol{G}$	Mass velocity
g	Gravitational acceleration
h	Convective heat transfer coefficient
h	Heat transfer coefficient
h_{lv}	Latent heat
K	Thermal conductivity
L	Length
L	Characteristic dimension; length
M_d	Masuda number
N	Number of channels
n_s	Number of fins
Nu	Nusselt number
P	Pressure
ΔP	Pressure drop
p	Pressure
p_f	Fin pitch normal to the fins
Q	Heat transfer rate
q	Heat flux
Re	Reynolds number
Re_l	Liquid Reynolds number

T	Temperature
T	Temperature
t	Time
U	Mean velocity
V	Volume
X	Effective quality
x	Vapor quality

Greek Symbols

α	Heat transfer coefficient; apex angle of the fin
β	Helix angle
δ	Liquid film thickness
Δp	Pressure drop
ΔT	Temperature difference
ε	Permittivity
μ	Dynamic viscosity
ρ	Density
σ	Surface tension

Subscripts

ave	Average
ev	Evaporation
in	Inlet
l	Liquid
s	Saturated
sub	Subcooled
v	Vapor

Abbreviations

CHF	Critical heat flux
CNT	Carbon nanotube
EHD	Electrohydrodynamic
ONB	Onset of nucleate boiling

Part I
Single Phase Flow

Abstract The advances in heat transfer enhancement techniques in all its aspects have been dealt with in this book. Part I deals with the single phase flow and Part II deals with the two-phase flow.
Industrial economy and growth are strongly dependent upon the performance of the machinery being used for different purposes. Irrespective of the type of industry, concepts of recycle and recovery are prevailing in order to save (i) energy and (ii) environment. Heat exchangers were devised and introduced in different plants in order to address energy saving and environmental issues. Needless to say that heat exchanger's performance needs to be improved as they consume major power in any plant. In this book source of major thermal resistance in heat exchangers is discussed in heat exchanger fundamental sections and the method of accessing their performance is explained from energy saving point of view. Since it is always desirable to improve the performance of existing heat exchanger, innovative methods proposed by different researchers are presented in a lucid manner. Size reduction without compromising thermohydraulic performance has always been prime objective of designers and in continuation of their effort compound techniques of heat transfer enhancement have evolved as a promising solution. The detailed discussion on this technique is presented in advances in compound technique section. Poor thermal conductivity has been identified one of the key factors responsible for reduced heat transfer. Recent advances pertaining to the technology of preparation and addition of high thermal conductivity nano particles in the heat transfer fluids and subsequent performance improvement are also discussed in the last section.

Keywords Heat transfer enhancement techniques • Active technique • Passive technique • Combined techniques • Heat transfer coefficient and friction factor correlations • Performance evaluation criteria

Chapter 1
Introduction

Abstract The advances in heat transfer enhancement techniques in all its aspects have been dealt with.

Keywords Heat transfer enhancement techniques • Active technique • Passive technique • Combined techniques • Heat transfer coefficient and friction factor correlations • Performance evaluation criteria

Energy saving is the challenge faced by the researchers and scientists. Heat exchange activities involved in food, process, chemical and automobile industries directly influence the economy of these industries. Many research projects have to focus on materials, machines and methods to ensure optimum utilization of energy. Enhanced heat transfer is one of the efforts to save energy. Heat transfer augmentation, in general, means an increase in heat transfer coefficient. Enhancement techniques can be classified either as passive, which require no direct application of external power, or as active, which require external power. Many permutation and combination of active and passive techniques have been adopted in an effort to get better performance than the individual technique. Such technique is named as compound technique of heat transfer augmentation. This part of the book will deal with single phase flow only. The references from which the materials are drawn in this part of the book are given below the text of Part I.

1.1 Heat Exchanger Fundamentals

The term heat exchanger applies to all equipment used to transfer heat between two streams (or two sources). Typical applications involve heating or cooling of a fluid stream and evaporation or condensation of single- or multicomponent fluid streams. In other applications, the objective may be to recover or reject heat, or sterilize, pasteurize, fractionate, distill, concentrate, crystallize, or control a process fluid. In a few heat exchangers, the fluids exchanging heat are in direct contact while in most of the heat exchangers, heat transfer between fluids takes place through a separating wall or into and out of a wall in a transient manner. In many heat exchangers, the fluids are separated by a heat transfer surface, and ideally they do not mix or leak. Such

S.K. Saha et al., *Advances in Heat Transfer Enhancement*,
DOI 10.1007/978-3-319-29480-3_1

exchangers are direct transfer type, or simply recuperators. In contrast, exchangers in which there is intermittent heat exchange between the hot and cold fluids—via thermal energy storage and release through the exchanger surface or matrix are indirect transfer type, or simply regenerators. Such exchangers usually have fluid leakage from one fluid stream to the other, due to pressure differences and matrix rotation/valve switching. Common examples of heat exchangers are shell-and tube exchangers, automobile radiators, condensers, evaporators, air preheaters, and cooling towers. If no phase change occurs in any of the fluids in the exchanger, it is sometimes referred to as a sensible heat exchanger. There could be internal thermal energy sources in the exchangers, such as in electric heaters and nuclear fuel elements. Combustion and chemical reaction may take place within the exchanger, such as in boilers, fired heaters, and fluidized-bed exchangers. For thorough understanding of heat exchangers with respect to their functions and the factors affecting performance, one needs to understand heat transfer fundamentals and principles involved in the working of heat exchangers. It is known that heat always flow from the higher temperature body to the lower temperature body by three primary modes named as conduction, convection, and radiation. In solids, heat transfer takes place through conduction mode while in fluids convection mode dominates. In a solid bar, if two surfaces are at different temperatures, the rate at which the heat flows is directly proportional to the temperature difference between the end surfaces, the cross-sectional area of the bar, a property of the solid called the "thermal conductivity" and is inversely proportional to the length. The equation of heat transfer by conduction is

$$Q_{conduction} = K.A.\Delta T / L$$

where Q is the rate of heat flow, K is the thermal conductivity of the material, L is the length, A is the cross-sectional area, and ΔT is the temperature difference between the hot and the cold ends of the solid. The conduction equation can be further written in terms of thermal resistance

$$Q_{conduction} = \Delta T / R_{conduction} \text{ where } R_{conduction} = L / K.A \text{ is the conductive thermal resistance.}$$

Convection is a very efficient means of heat transfer. Unlike conduction, where the molecules are stationary, in convection, the molecules are moving. Because the molecules are moving, the rate of heat transfer can be considerably higher than in conduction. The equation for heat transfer by convection is

$$Q_{convection} = h.A.\left(T_{solid} - T_{fluid}\right)$$

Where A is the area where the solid surface and the fluid interact, T_{solid} and T_{fluid} are the temperatures of the solid surface and the fluid respectively, and h is the film coefficient. The film coefficient varies widely depending on the properties of the fluid and how fast it is moving.

The convection equation can be rewritten in terms of thermal resistance as

$$Q_{convection} = \left(T_{solid} - T_{fluid}\right) / R_{convective} \text{ where } R_{convective} = 1 / h.A \text{ is the convective thermal resistance.}$$

From the above discussion, it is clear that as the heat flow takes place due to temperature difference, it experiences resistance, called thermal resistance. Now since in heat exchangers, solid as well as fluid media are involved for the purpose of heat exchange, identifying the sources of thermal resistance for the scope of improving heat transfer performance is vital.

Scopes of improvement of thermal conductivity of solid materials that can be used in heat exchangers are limited to the development of new high conductivity materials hence reduction in conductive thermal resistance is limited and therefore majority of research works related to heat exchanger performance improvement focus on reducing the convective thermal resistance in heat exchangers for heat transfer augmentation.

1.2 Impact of Boundary Layer Thickness on Thermal Resistance

As a fluid moves past a stationary surface, there will be a velocity gradient in the fluid stream. The slowest moving particles will be the ones that are closer to the wall and are being slowed down by friction with the surface, while the fastest moving particles will be the ones that are farther away.

A useful engineering approximation is to assume that there is a thin layer of fluid which is completely stationery along the surface. This thin layer is called a boundary layer as shown in Fig. 1.1. Since the boundary layer is stationary, the heat transfer through this layer is determined using thermal conductivity equations instead of convection equations. Thermal performance is affected by the thickness of the boundary layers in a heat exchanger. The larger the thickness of the boundary layers, the larger will be the thermal resistance. When the fluid velocity increases, the boundary layers become smaller. This has the effect of increasing the film coefficient and thereby reducing the thermal resistance.

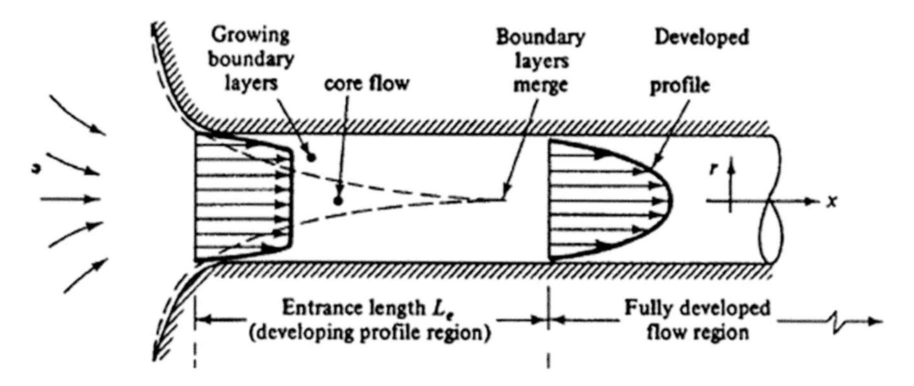

Fig. 1.1 Concept of boundary layer in internal flow

1.3 Influence of Types of Flow on Heat Transfer

Laminar flow offers greater challenge to heat transfer augmentation than the turbulent flow because

(i) Boundary layer thickness of laminar flow is more compared to the turbulent boundary layer
(ii) Laminar flow being stream line flow it lacks intermixing of the fluid particles, resulting in less heat exchange potential.

On the other hand in turbulent flow the boundary layer is confined to thin layer in the vicinity of the tube wall. Also fluid particles readily diffuse into the adjacent layer particle, promoting heat transfer.

1.4 Advantages and Disadvantages of Laminar Flow

Mostly nature of fluid itself causes the flow to become laminar and this situation often arises where the fluid is highly viscous such as in paint industries, food industries, oil industries etc. On the other hand, sometimes situation of laminar flow is created intentionally in order to get the advantages of entrance length (a region of relatively higher heat transfer co-efficient) and reduction in consumption of pumping power. For example, in compact heat exchangers, the fluid flow is intentionally laminar because of the reduction in hydraulic diameter and if compact heat exchangers are not operated at low Reynolds number, there will be considerable increase in frictional losses. Moreover compact heat exchangers also get advantage of entrance effect due to reduced length of the tube.

Chapter 2
Meaning of Improved Heat Exchanger

Abstract The advances in heat transfer enhancement techniques in all its aspects have been dealt with.

Keywords Heat transfer enhancement techniques • Active technique • Passive technique • Combined techniques • Heat transfer coefficient and friction factor correlations • Performance evaluation criteria

The performance of heat exchanger is defined by its thermal performance (i.e. by improvement in convective heat transfer coefficient) and by its hydraulic performance (i.e. amount of power consumed in pumping the fluid). Hence the term thermohydraulic performance is used as an indicator to heat exchanger performance. By performance evaluation we mean the comparison of heat transfer and pressure drop characteristics of an enhancement device with a reference device which possesses the same basic dimensions except the part of enhancement structure. Different methods (active/passive/compound) are used for enhancing the heat transfer but the one which leads to larger heat transfer enhancement with relatively less increment in pumping power penalty will be an acceptable solution. Thus with the development of heat transfer enhancement techniques, the question of how to evaluate the effectiveness of enhancement technique also needs to be addressed. A qualitative comparison for enhanced heat transfer surface is an important issue in evaluating the performance improvement of the enhancement techniques. Broadly speaking, all the existing assessment methods, or criteria, can be classified into two categories. In one category of evaluation methods, the 2nd law of the thermodynamics is applied and either entropy or exergy generated is determined, while in the other category, discussion is only based on the 1st law of thermodynamics. Factors like ease of manufacturing, safety, reliability etc. also need to be incorporated in evaluation of the technique. But it seems intangible to propose a performance evaluation technique that will include all the said factors. Nevertheless, in the present scenario, the purpose of energy-saving of the heat transfer enhancement has received more attention of the international heat transfer community. Keeping energy saving as the prime purpose in mind the evaluation approach based on the comparison between enhanced and original (or reference) heat transfer surfaces (or structure) for identical flow rate, identical pressure drop and identical pumping power seems more logical. In this regard the performance evaluation plot of enhanced heat transfer

S.K. Saha et al., *Advances in Heat Transfer Enhancement*,
DOI 10.1007/978-3-319-29480-3_2

techniques from energy saving point of view is important. The proposed plot should be capable of displaying

(i) Whether the enhancement technique under investigation can really save energy (pumping power).
(ii) Comparison of effectiveness of enhancement techniques from energy saving point of view
(iii) Which enhancement technique (among the available ones) will be more suitable in a prevailing situation (i.e. we can select the better heat transfer technology through corresponding performance evaluations).

Experimental and numerical investigations give the characteristics of heat transfer and pressure drop of an enhanced surface and those of the reference one. The correlations of friction factor and heat transfer for the reference surface and the enhanced surface help evaluate the performance comparison between the two surfaces.

2.1 Derivation of Basic Equations for Performance Evaluation and Performance Plot

Derivation of basic equations is done assuming that

1. The thermo-physical properties of fluid are constant.
2. Heat transfer area used for calculating the convective heat transfer coefficient of enhanced surface is the same as that of the reference one.
3. The cross-sectional area used for calculating the average velocity of fluid of enhanced surface is the same as that of the reference one.
4. The reference dimension used for calculating the dimensionless characteristic number of enhanced surface is the same as that of the reference one.

For the reference surface the average friction factors and Nusselt number can be fitted as

$$f_o(Re) = C_1 Re^{m_1} \tag{2.1}$$

$$Nu_o(Re) = C_2 Re^{m_2} \tag{2.2}$$

Ratio of friction factor of an enhanced surface over the reference surface at same Reynolds number (Re) will be

$$\frac{f_e(Re)}{f_o(Re)} = \left(\frac{f_e}{f_o}\right)_{at\ same\ Re} \tag{2.3}$$

Similarly, Ratio of Nusselt number of an enhanced surface over the reference surface at same Reynolds number (Re) will be

$$\frac{Nu_e(Re)}{Nu_o(Re)} = \left(\frac{Nu_e}{Nu_o}\right)_{at\ same\ Re} \tag{2.4}$$

Further, at different Reynolds numbers, the ratios of Nusselt number and friction factor of the enhanced heat transfer surface over the reference one is expressed by

$$\left(\frac{f_e}{f_o}\right)_{at\ different\ Re} = \frac{f_e(Re)}{f_o(Re_o)} \tag{2.5}$$

$$\frac{Nu_e(Re)}{Nu_o(Re_o)} = \left(\frac{Nu_e}{Nu_o}\right)_{at\ different\ Re} \tag{2.6}$$

From Eqs. (2.1) and (2.3) we get

$$f_e(Re) = C_1 Re^{m_1}\left(\frac{f_e}{f_o}\right)_{at\ same\ Re} \tag{2.7}$$

$$\text{But} \quad f_o(Re_o) = C_1 Re_o^{\ m_1} \tag{2.8}$$

Now, substituting the values of $f_e(Re)$ and $f_o(Re_o)$ from Eqs. (2.7) and (2.8) in to the Eq. (2.5) we get an expression of friction factor ratio at different Reynolds numbers as

$$\left(\frac{f_e}{f_o}\right)_{at\ different\ Re} = \frac{f_e(Re)}{f_o(Re_o)} = \left(\frac{Re}{Re_o}\right)^{m_1}\left(\frac{f_e}{f_o}\right)_{at\ same\ Re} \tag{2.9}$$

Similarly, we can get the ratio of the Nusselt number at different Reynolds number:

$$\left(\frac{Nu_e}{Nu_o}\right)_{at\ different\ Re} = \frac{Nu_e(Re)}{Nu_o(Re_o)} = \left(\frac{Nu_e}{Nu_o}\right)_{at\ same\ Re}\left(\frac{Re}{Re_o}\right)^{m_2} \tag{2.10}$$

It is clear from Eqs. (2.9) and (2.10) that once the basic correlations of the reference surface and the ratios of friction factors (and Nusselt numbers) of enhanced and reference structures at the same Reynolds number are available, then the two ratios at different Reynolds number can be calculated.

In the performance evaluation of enhanced structure, the ratio of heat transfer rate is of great importance. Hence the ratio of heat transfer rate of the enhanced and reference surfaces can be presented form basic definition of rate of heat transfer as

$$\left(\frac{Q_e}{Q_o}\right) = \frac{(h.A.\Delta t)_e}{(h.A.\Delta t)_o} = \frac{\left(Nu.\frac{K_f}{D}.\ A.\ \Delta t\right)_e}{\left(Nu.\ \frac{K_f}{D}.\ A.\ \Delta t\right)_o} \tag{2.11}$$

But when the heat transfer rates of two surfaces are compared for surface evaluation, it is reasonable to assume the same temperature difference. Also, following the assumptions, we have

$$\frac{\left(\frac{K_f}{D}.\ A.\ \Delta t\right)_e}{\left(\frac{K_f}{D}.\ A.\ \Delta t\right)_o} = 1 \tag{2.12}$$

Substituting Eq. (2.12) into Eq. (2.11), the ratio of heat transfer rate at different Reynolds no. becomes

$$\left(\frac{Q_e}{Q_o}\right) = \left(\frac{Nu_e}{Nu_o}\right) \tag{2.13}$$

Now, expressions of heat transfer ratio under the three constraints (i.e. for identical pumping power, identical pressure drop and for identical flow rate) are derived.

The ratio of power consumption of enhanced and reference surfaces can be presented as

$$\left(\frac{P_e}{P_o}\right) = \frac{(A_c.\ V.\ \Delta p)_e}{(A_c.\ V.\ \Delta p)_o} = \frac{\left(A_c.\ V.\ f.\ L.\ \rho.V^2/D\right)_e}{\left(A_c.\ V.\ f.\ L.\ \rho.V^2/D\right)_o} \tag{2.14}$$

But according to assumptions made,

$$\frac{\left(A_c.\ L.\ \rho/D\right)_e}{\left(A_c.\ L.\ \rho/D\ \right)_o} = 1$$

Substituting this in Eq. (2.14) we get

$$\left(\frac{P_e}{P_o}\right) = \frac{\left(f.\ V^3\ \right)_e}{\left(f.\ V^3\ \right)_o} = \frac{f_e}{f_o}\frac{\left(\ Re^3\ \right)_e}{\left(Re^3\ \right)_o} \tag{2.15}$$

Now for same power consumption condition,

$$\left(\frac{P_e}{P_o}\right) = 1$$

Substituting this in Eq. (2.15) we get

$$\frac{f_e}{f_o} = \left(\frac{Re}{Re_o}\right)^{-3} \tag{2.16}$$

Substituting Eq. (2.16) in Eq. (2.9) we get

$$\left(\frac{Re}{Re_o}\right)^{-3}=\left(\frac{Re}{Re_o}\right)^{m_1}\left(\frac{f_e}{f_o}\right)_{at\ same\ Re}$$

Simplifying we get,

$$\frac{Re}{Re_o}=1/\left(\frac{f_e}{f_o}\right)_{Re}^{\frac{1}{3+m_1}} \tag{2.17}$$

Substituting Eq. (2.17) in Eq. (2.10) we get heat transfer ratio under the **constraint of identical power consumption** as,

$$\left(\frac{Q_e}{Q_o}\right)=\left(\frac{Nu_e}{Nu_o}\right)_{at\ diff.\ Re}=\left(\frac{Nu_e}{Nu_o}\right)_{at\ same\ Re}/\left(\frac{f_e}{f_o}\right)_{Re}^{\frac{m_2}{3+m_1}} \tag{2.18}$$

Similar formulation for the heat transfer ratio can be derived for the constraint of identical pressure drop. The ratio of pressure drop of the enhanced and the reference surfaces is given by

$$\frac{\Delta p_e}{\Delta p_o}=\frac{\left(f.\ L.\ \rho.V^2/D\right)_e}{\left(f.\ L.\ \rho.V^2/D\right)_o} \tag{2.19}$$

But according to assumptions, we have

$$\frac{\left(A_c.\ L.\ \rho/D\right)_e}{\left(A_c.\ L.\ \rho/D\right)_o}=1$$

Substituting in Eq. (2.19) we get,

$$\frac{\Delta p_e}{\Delta p_o}=\frac{\left(f.\ V^2\right)_e}{\left(f.\ V^2\right)_o}=\frac{f_e}{f_o}\left(\frac{Re}{Re_o}\right)^2 \tag{2.20}$$

Now, for identical pressure drop, $\frac{\Delta p_e}{\Delta p_o}=1$

Substituting this in Eq. (2.20) we get

$$\frac{f_e}{f_o}=\left(\frac{Re}{Re_o}\right)^{-2} \tag{2.21}$$

Substituting Eq. (2.21) in Eq. (2.9) we get

$$\left(\frac{Re}{Re_o}\right)^{-2}=\left(\frac{Re}{Re_o}\right)^{m_1}\left(\frac{f_e}{f_o}\right)_{at\ same\ Re}$$

Simplifying we get,

$$\frac{Re}{Re_o} = 1/\left(\frac{f_e}{f_o}\right)_{Re}^{\frac{1}{2+m_1}} \tag{2.22}$$

Substituting Eq. (2.22) in Eq. (2.10) we get heat transfer ratio under the **constraint of identical pressure drop** as,

$$\left(\frac{Q_e}{Q_o}\right) = \left(\frac{Nu_e}{Nu_o}\right)_{at\ diff.\ Re} = \left(\frac{Nu_e}{Nu_o}\right)_{at\ same\ Re} / \left(\frac{f_e}{f_o}\right)_{Re}^{\frac{m_2}{2+m_1}} \tag{2.23}$$

Finally, expression of heat transfer ratio **under the constraint of identical flow rate** can be found by substituting $\frac{Re}{Re_o} = 1$ (which is the condition of identical flow rate) in Eq. (2.10) we get

$$\left(\frac{Q_e}{Q_o}\right) = \left(\frac{Nu_e}{Nu_o}\right)_{at\ same\ Re} \tag{2.24}$$

Now, three equations can be collectively written as

$$C_i = \left(\frac{Nu_e}{Nu_o}\right)_{Re} / \left(\frac{f_e}{f_o}\right)_{Re}^{k_i} \quad (i = P, \Delta p, V) \tag{2.25}$$

Where P, Δp, V stand for identical pumping power, identical pressure drop and identical flow rate, respectively.

Taking logarithm

$$\ln\left(\frac{Nu_e}{Nu_o}\right)_{Re} = \ln C_i + K_i \ln\left(\frac{f_e}{f_o}\right)_{Re} \tag{2.26}$$

2.2 Plot Interpretation

Clearly, Eq. (2.26) represents equation of straight line in log–log coordinate system, where K_i is the slope and $\ln C_i$ is the intercept on y- axis. The physical interpretation of $\ln C_i$ and K_i are as follows.

The value of $\ln C_i$ is an indication of heat transfer enhancement under different constraint conditions. When $\ln C_i = 0$, the straight line crosses the point of coordinate (1,1) and this means that the enhanced and reference surfaces possess the same heat transfer rates under the corresponding constraint conditions. When $\ln C_i > 0$, the enhanced heat transfer surface can transfer a larger heat transfer rate than that of the reference one, and when $\ln C_i < 0$, the enhanced heat transfer surface transmits a lower heat transfer rate.

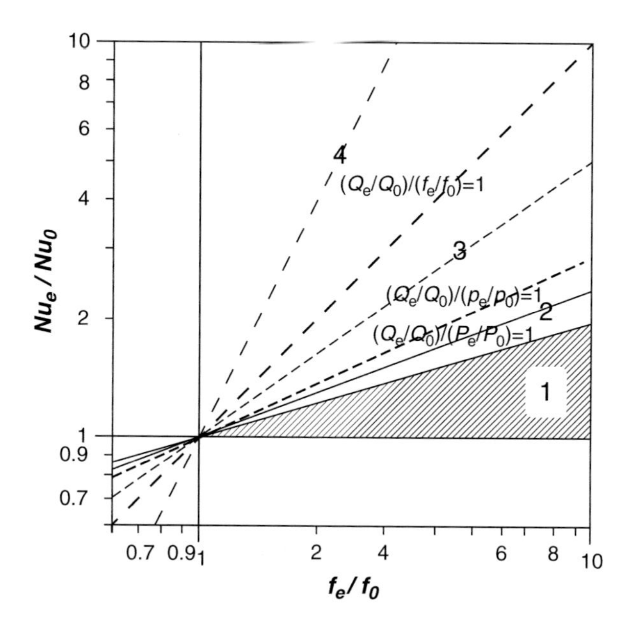

Fig. 2.1 A performance evaluation plot for enhancement technique

The performance plot as shown in Fig. 2.1 is divided into four regions. Region 1 is characterized by enhanced heat transfer without energy-saving, where the enhancement of heat transfer rate is less than the increase of power consumption. Region 2 is featured by enhanced heat transfer with the same pump power consumption, i.e., where the enhanced surface presents higher heat transfer rate than reference one under the same pumping power consumption. In Region 3 enhanced heat transfer can be obtained with the same pressure drop, i.e., where the enhanced surface presents higher heat transfer rate than the reference one under the identical pressure drop constraint. Region 4 is the most desirable one where the augmentation of heat transfer rate is more than the increase in friction coefficient under the same flow rate. Lines forming Regions are drawn using Eq. (2.25) under different constraint conditions and the values of variables m_1 and m_2 come from the relevant correlations.

2.3 How to Use the Proposed Performance Plot

In the proposed plot, the lines are representing the performance of some enhancing structure and are named as working lines. Any point on the working line is called a working point. There are three kinds of working lines corresponding to the three constraints. The shaded region is the case of enhanced heat transfer without energy-saving, i.e., when a working point of an enhancement technique is located in this region the consumption of unit pumping power leads to less heat transfer rate compared with that of the reference one. Thus for energy-saving purposes, the working point should be located outside the shaded region. Further, for energy-saving purposes, the larger the slope of the working line, the better the enhanced technique. As far as different points on the same working line are concerned, they have the same enhanced heat transfer ratio under corresponding friction loss constraint. Therefore, the working lines are the contours of the heat transfer enhancement ratio under different constraints.

Chapter 3
Advances in Passive Techniques

Abstract The advances in heat transfer enhancement techniques in all its aspects have been dealt with.

Keywords Heat transfer enhancement techniques • Active technique • Passive technique • Combined techniques • Heat transfer coefficient and friction factor correlations • Performance evaluation criteria

According to the technological development achieved, the technology in applied heat transfer may be classified into "generations". The first generation is represented by the bare tube, the second by plain fins, third by longitudinal vortex generators on fins and the fourth by compound enhancement techniques. Some important advances in the domain of heat transfer enhancement are discussed below.

3.1 Effort to Improve Existing Heat Exchangers

Usage of insert devices is still receiving attention of researchers for heat transfer augmentation for the following reasons

(i) Insert devices give scope of improving the performance of existing heat exchangers
(ii) Easy removal for inspection /maintenance/replacement is possible.

Many modifications have been carried out in the last 10 years in the insert devices like twisted tapes. Experimental investigations pertaining to the advances in these insert devices are being discussed here. Insert devices are commonly used as enhancement tool in marine applications and chemical industries where scaling or fouling problem is dominating also in some specific applications such as heat exchangers dealing with fluids of low thermal conductivity. In general, insert devices are placed in the flow passage and this reduces the hydraulic diameter of the flow passage. Heat transfer enhancement in a tube flow by inserts such as twisted tapes, screw tape is mainly due to flow blockage, partitioning of the flow and secondary flow. Flow blockage increases the pressure drop and leads to increased viscous effects because of a reduced free flow area. Blockage also increases the flow

S.K. Saha et al., *Advances in Heat Transfer Enhancement*,
DOI 10.1007/978-3-319-29480-3_3

velocity and in some situations leads to a significant secondary flow. Secondary flow further provides a better thermal contact between the surface and the fluid because secondary flow creates swirl and the resulting mixing of fluid improves the temperature gradient, which ultimately leads to a high heat transfer coefficient.

3.1.1 Investigation of Heat Transfer Enhancement by Perforated Helical Twisted-Tapes

Though success has been observed when the twisted tapes are being used in heat exchangers for heat transfer augmentation, yet increased pumping power with the usage of insert devices remains to be a concern. Investigation has been made to study the prospect of using perforated helical twisted tape with a view to reduce pumping penalty. The tape was fabricated using Aluminum sheets with an axial-length of 1500 mm (L), a width of 12.6 mm (w) and a thickness of 0.8 mm (δ). Figure 3.1 shows the important geometrical parameters and notations used for perforated helical twisted tapes. The studied parameters include perforation diameter ratio (d/w) and perforation pitch ratio (s/w).

Perforated helical twisted-tape (P-HTTs) were prepared at three different ratios of perforation diameter to tape width, d/w = 0.2, 0.4 and 0.6, and three different ratios of perforation pitch to tape width, s/w = 1, 1.5 and 2 while a helical pitch ratio and a twist ratio were fixed at P/D = 2 and y/w = 3. These configurations are shown in Figs. 3.2 and 3.3.

The experiment was conducted under uniform heat flux condition and the Reynolds number was varied from 6000 to 20,000. The thermohydraulic characteristics have been presented in terms of Nusselt number, friction factor and thermal performance factor. Figure 3.4 shows the variation of Nusselt number with Reynolds number for the plain tube and the tubes equipped with P-HTTs and the typical helical twisted-tape (HTT). Following conclusions may be inferred from the said plot.

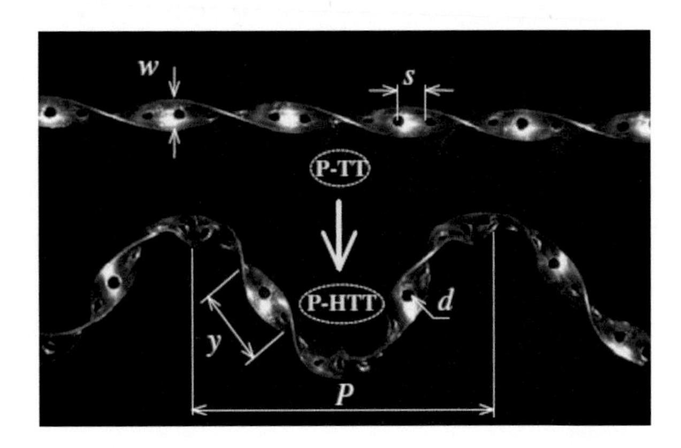

Fig. 3.1 Geometrical parameters of perforated helical twisted tape

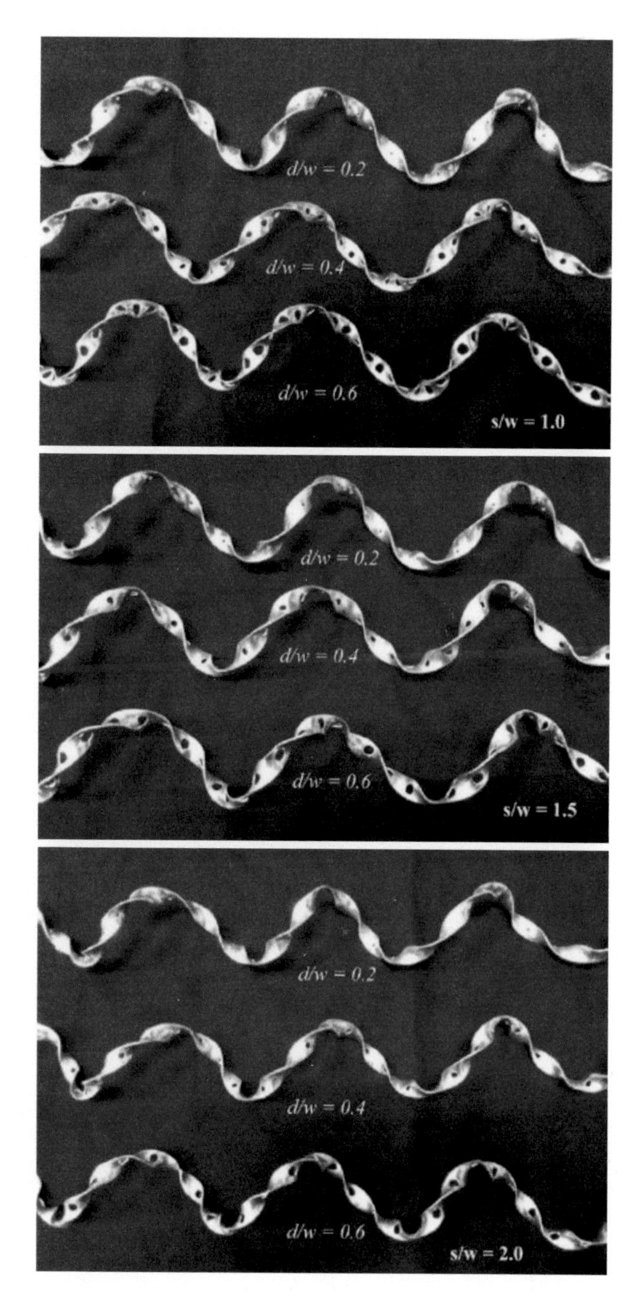

Fig. 3.2 Geometry details of P-HTTs with different perforation diameter ratios

(i) Nusselt number shows an increasing trend and at a given Reynolds number, the use of twisted tape inserts leads to considerable increase in Nusselt number as compared to that of the plain tube. The Nusselt numbers for the tubes with tape inserts are enhanced between 51.4 and 103.8 %, over that for the plain tube.

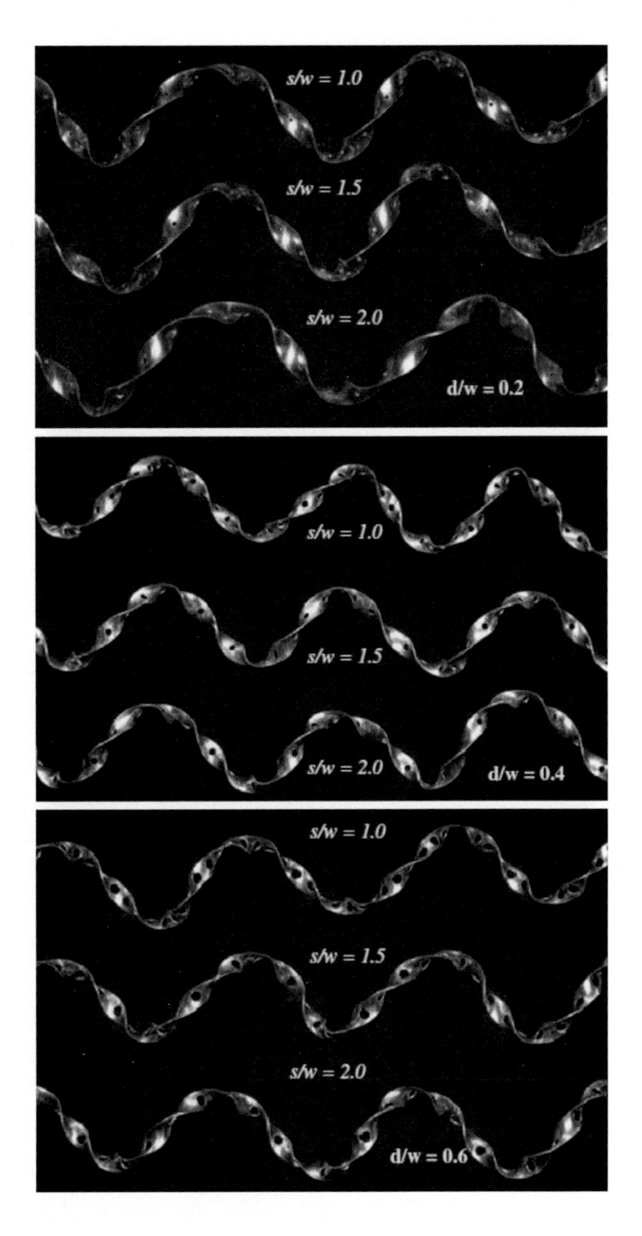

Fig. 3.3 Geometry details of P-HTTs with different perforation pitch ratios

(ii) Slightly Higher heat transfer augmentation (1.1–13.9 %) is achieved with non-perforated helical twisted tape than the perforated helical twisted tape. This is because perforation in the tape results in reduced intensity of secondary flow penetrating the boundary layer.

However, when Nusselt number ratio (Nu_t/Nu_p) variation were plotted, as shown in Fig. 3.5, it was observed that Nu_t/Nu_p ratio increases as Reynolds number decreases. This is because the thermal boundary becomes thicker as Reynolds

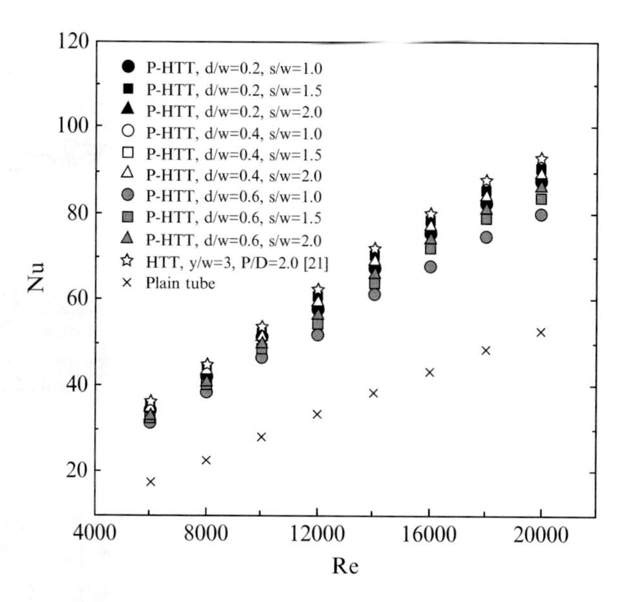

Fig. 3.4 Variations of Nu with Reynolds number

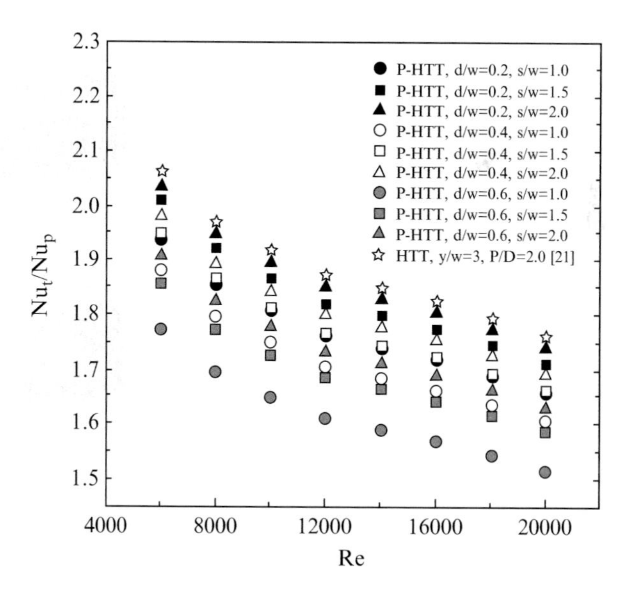

Fig. 3.5 Variations of Nu_t/Nu_p with Reynolds number

number decreases, thus the effect of boundary layer destruction by inserts turns out to be more prominent. Further, it is also found that Nusselt number increases with decreasing perforation diameter ratio (d/w) and increasing perforation pitch ratio (s/w). The results indicate that the P-HTT with smaller perforation diameter (d) and larger perforation pitch which is more similar to the HTT offers more efficient heat transfer enhancement as it possesses a larger surface for flow disturbance (or a smaller perforated area), leading to a higher turbulence intensity imparted to the flow between the P-HTT and tube wall. The P-HTTs with d/w = 0.2 yield 2.9–3.2 % and 6.8–9.3 % higher Nusselt numbers than the ones with d/w = 0.4 and d/w = 0.6, respectively while P-HTTs with s/w = 2.0 yield 5.2–7.6 % and 1.5–2.8 % higher Nusselt numbers than the ones with s/w = 1.0 and 1.5, respectively.

3.1.1.1 Observation About Pressure Loss

Figures 3.6 and 3.7 depict the variation of friction factor with Reynolds number for the plain tube without any insert, tubes with perforated helical twisted tapes and with helical twisted tape. As expected, tube with perforated helical twisted tape experienced around 2.6–16.4 % lower friction factors than that in case with non-perforated helical twisted tape. This lower penalty of pumping power is due to the decrease in a fluid flow blockage and a flowing path length as well as the forces exerted by reverse flow impinging on the tube wall. Also, reduction in friction factor was observed with increasing perforation diameter ratio (d/w) and decreasing perforation pitch ratio (s/w), as the perforated area increases and thus a fluid flow blockage diminishes.

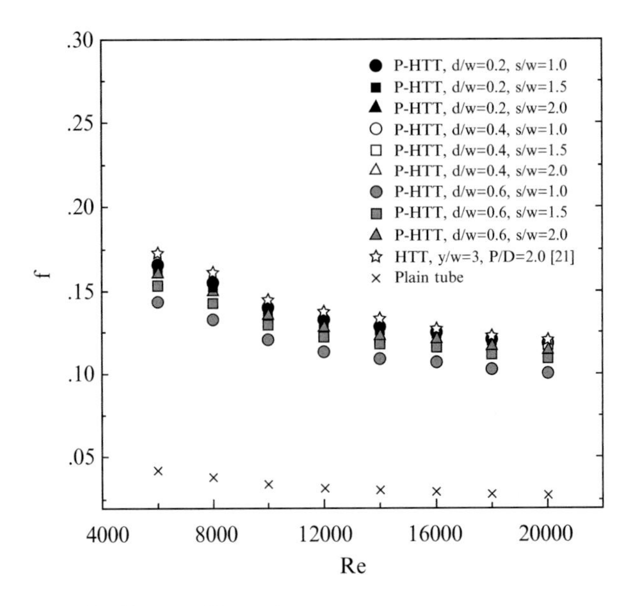

Fig. 3.6 Variations of f with Reynolds number

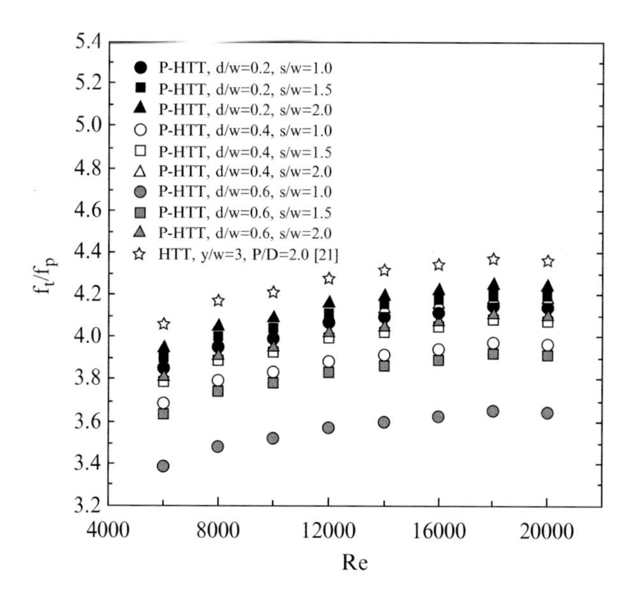

Fig. 3.7 Variations of f_t/f_p with Reynolds number

3.1.1.2 Comparing Thermal Performance Factor

Variation in thermal performance factor with Reynolds number is shown in the Fig. 3.8. Comparison has been done for the tubes equipped with perforated-helical twisted tapes and helical twisted Tape as well as the plain tube. It is found that the usage of perforated helical twisted tape though results in reduction in frictional losses compared to the non-perforated helical twisted tape, it yields lower thermal performance factor (0.5–9.5 %) than the helical twisted tape. It is further noticed that at a given Reynolds number, thermal performance factor increases as perforation diameter ratio (d/w) decreases and perforation pitch ratio (s/w) increases. The maximum thermal performance factor of 1.28 is obtained by using the P-HTT with the smallest perforation diameter (d/w = 0.2) and the largest perforation pitch (s/w = 2.0) at the lowest Reynolds number of 6000.

3.1.2 V-Cut Twisted Tape Insert

Twisted tape may be modified by introducing V-cut in the plain twisted tape on both top and bottom alternately in the peripheral region with different dimensions of depth and width to improve the fluid mixing near the walls of the test section. The Heat transfer and pressure drop characteristics in a circular tube fitted with and without V-cut twisted tape insert have been studied. Figures 3.9 and 3.10 depict the plain twisted tapes of different twist ratios and the modified twisted tape with V-cut having different depth ratios (DR) (d_e/W) and width ratios (WR) (w/W).

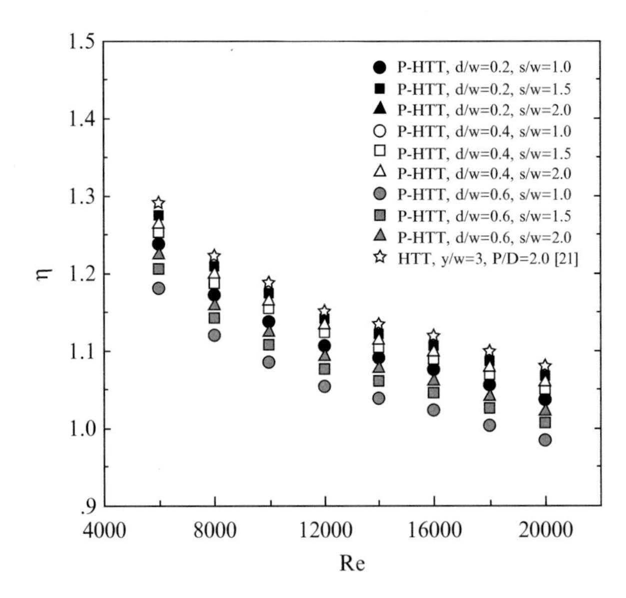

Fig. 3.8 Variation of thermal performance factor with Reynolds number

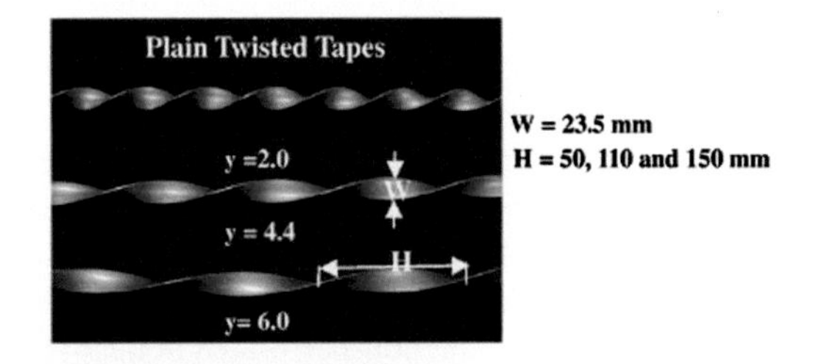

Fig. 3.9 Geometries plain twisted tapes

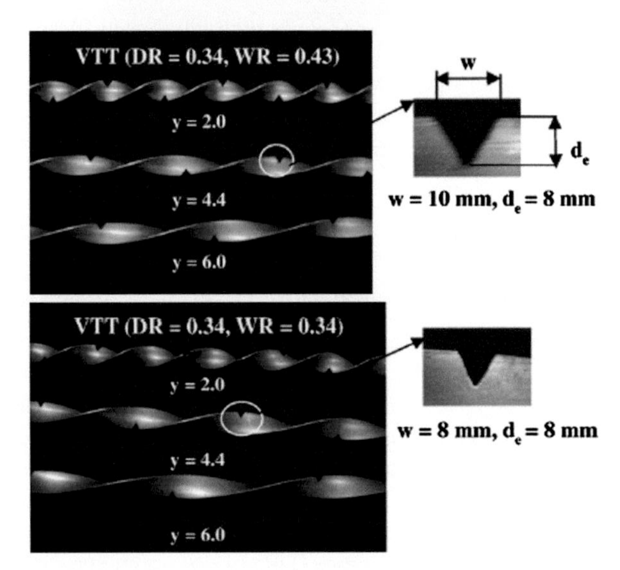

Fig. 3.10 Geometries V-cut twisted tapes

3.1.2.1 Influence of V-Cut and Its Geometrical Parameters on Heat Transfer

Figure 3.11 depicts the variation of Nusselt number with Reynolds number for V-cut twisted tape, plain twisted tape and plain tube. Clearly, Nusselt number is increasing with increasing Reynolds number and V-cut twisted tapes have intensified heat transfer 1.36–2.46 times more than the plain twisted tape. This enhancement may be attributed to the turbulence induced in the proximity of the wall and the vortex motion induced behind the V-cut. Further, it was observed that geometrical parameters of V-cut have significant influence on heat transfer augmentation. The heat transfer increases with increasing depth ratios and decreasing width ratios. This is because more vortices are generated behind the cut as depth ratio increases and width ratio decreases, resulting in improved heat transfer.

3.1.2.2 Frictional Loss Aspect

Friction factor shows an increasing trend with increasing Reynolds number as shown in Fig. 3.12. For V-cut twisted tape, 1.05–1.59 times increase from plain twisted tape values has been observed. Also, friction factor increase with increasing depth ratios and decreasing width ratios has been observed. This is because turbulence is intensified as the depth ratio increases and width ratio decreases, resulting in increased frictional loss.

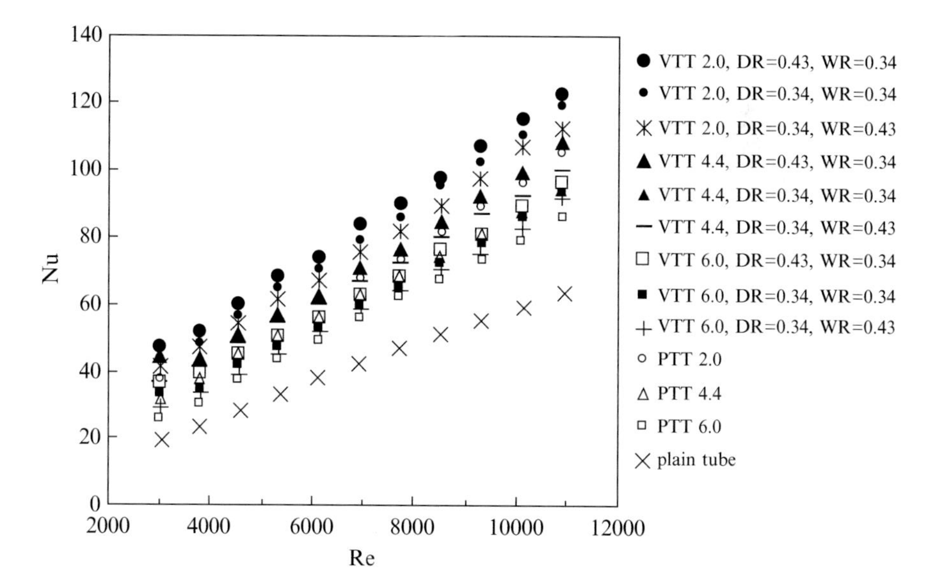

Fig. 3.11 Nusselt number vs Reynolds number

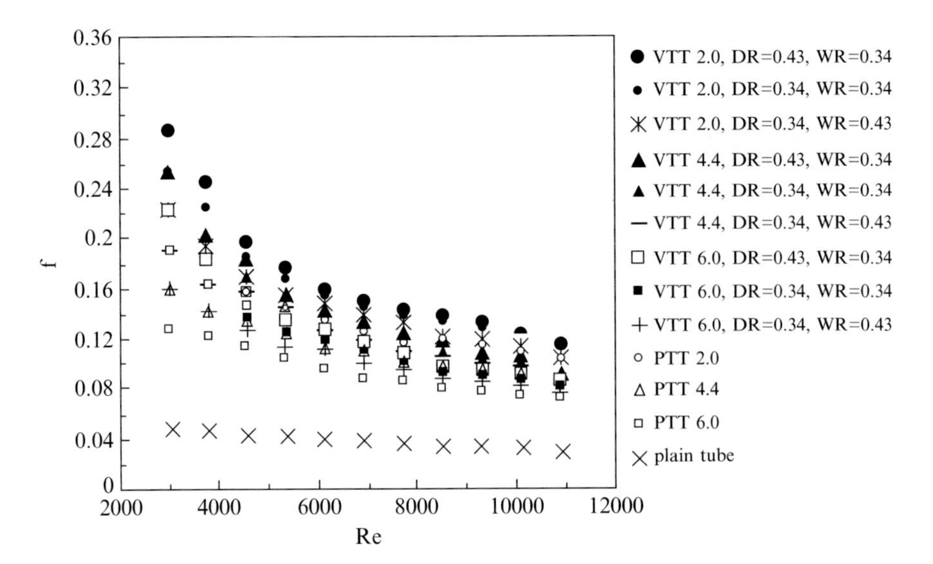

Fig. 3.12 Friction factor vs Reynolds number

3.2 Effect of Internally Grooved Shape on Heat Transfer Augmentation

The effect of different geometric groove shapes (circular, trapezoidal and rectangular) on heat transfer and friction characteristics of a fully developed turbulent air flow has been studied in Reynolds number range 10,000–38,000. During the experiment, depth and length of all grooves were fixed as 3 and 6 mm, respectively and the same number of grooves (n = 99) was obtained for both circular and rectangular grooved pipes, while the trapezoidal grooved pipe has less number of grooves (n = 79) because of the fixed groove length.

Figure 3.13 shows different geometric shapes of the grooved tubes used for experimental study.

The effect of groove shapes on heat transfer rate is compared with the plain tube and is shown in Fig. 3.14.

Following observations have been reported.

1. For all the cases of groove, (irrespective of their shape) heat transfer has increased compared to the plain tube.
2. The ratio of the Nusselt number first increases sharply up to Re = 26,000 and then onwards it becomes almost constant, for all cases of groove.
3. The ratio of Nusselt number for trapezoidal and circular groove is (1.35–1.63 for circular and 1.33–1.58 for trapezoidal groove) nearly the same while the ratio varies from 1.38–1.47 for rectangular groove.
4. Friction coefficients of all grooved pipes have nearly same values; irrespective of Reynolds number while friction coefficient decreases gradually with increasing Reynolds number for smooth tube (Fig. 3.15).

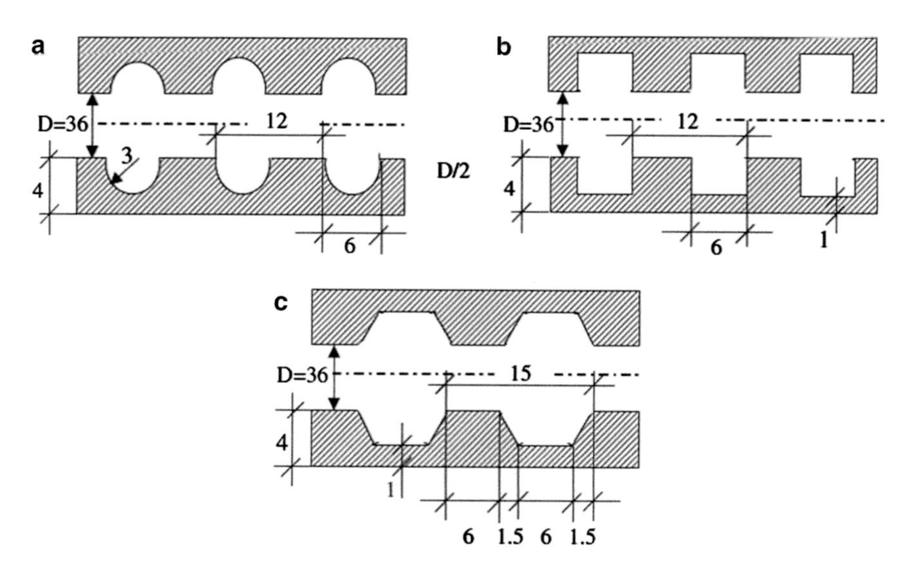

Fig. 3.13 The geometric shapes of the grooved tube, dimensions in mm (**a**) circular, (**b**) rectangular and (**c**) trapezoidal grooves

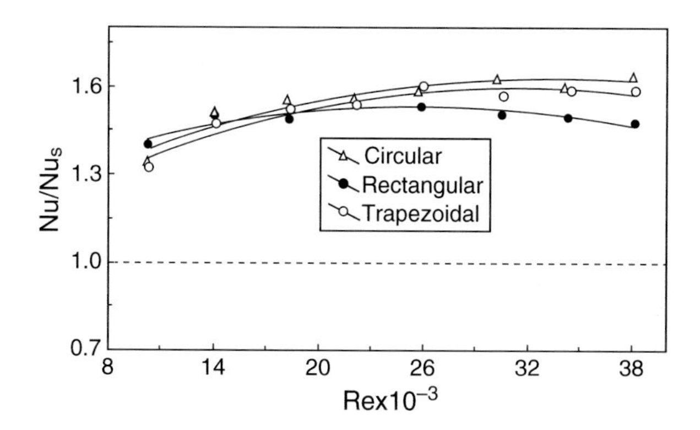

Fig. 3.14 Variation of Nusselt number ratio with Reynolds number for different grooved tubes

Following co-relations have been proposed for the grooved surfaces:

For circular grooved tube,
$Nu = 0.0148\ Re^{0.889}\ Pr^{1/3}$
$f = 0.0356\ Re^{0.124}$
for rectangular grooved tube,
$Nu = 0.0339\ Re^{0.803}\ Pr^{1/3}$
$f = 0.071\ Re^{0.062}$
for trapezoidal grooved tube,
$Nu = 0.014\ Re^{0.893}\ Pr^{1/3}$
$f = 0.0428\ Re^{0.107}$

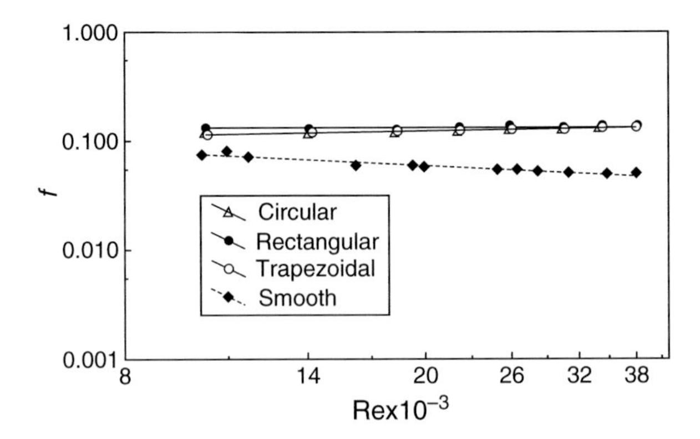

Fig. 3.15 Variation of friction factor with Reynolds number for smooth and different grooved tubes

3.3 Twisted Elliptical Tubes

This type of modified tube finds application in shell and tube type heat exchangers with an objective to reduce its size. Experimental investigations have been made to study the effect of aspect ratio and twist pitch on thermo-hydraulic performance, using water as the working fluid and varying the Reynolds number from around 600 to 55,000, covering laminar, transition and turbulent flow regimes. The geometry of elliptical tube is shown in Fig. 3.16.

Where S is 360° twist pitch while dimensions A and B stand for major and minor dimensions of the cross section. The experiment was conducted using a, smooth tube and five twisted tubes as shown in Fig. 3.17 while their geometrical specifications are shown in the Table 3.1.

3.3.1 Heat Transfer Performance

Nusselt number, a dimensionless parameter, is used as an indicator of thermal performance. Figure 3.18 depicts variation of Nusselt numbers with Reynolds numbers for the twisted elliptical tubes and smooth tube. It is observed that (1) for all the tubes, Nusselt number increases as the Reynolds number increases. (2) All the twisted elliptical tubes (TETs) performed better than the plain tube and the twisted tube number 1 performed the best among the five twisted tubes. (3) The effect of the twist pitch on the heat transfer performance of a twisted elliptical tube is more notable than that of the tube aspect ratio.

The effectiveness of modified tube is estimated by plotting ratio of the Nusselt number for the enhanced tube (Nu_e) to that for the plain tube (Nu_s) as shown in Fig. 3.19.

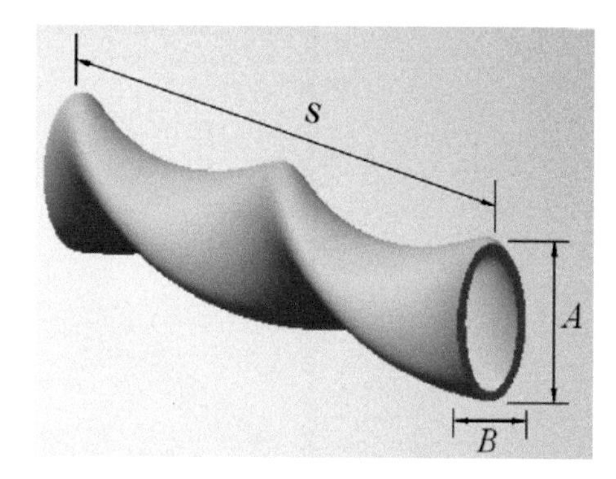

Fig. 3.16 Geometry of twisted tube

Fig. 3.17 Twisted elliptical tubes

Table 3.1 Specifications of twisted elliptical tubes

Tube no.	Twist pitch, S (mm)	Aspect ratio, A/B
1	104	1.60
2	152	1.90
3	192	2.15
4	192	1.76
5	192	1.49

It is observed that first Nusselt number ratios for all the tested TETs tend to increase up to Reynolds number 2300 and then it shows a decreasing trend with the rise of Reynolds number. From this trend, it may be concluded that the heat transfer enhancement of the TETs is more significant at lower fluid velocities corresponding to the laminar and transition flow regime. The Nusselt number ratios for TET No. 1 that has the best heat transfer performance are around 3.9–4.8 for Reynolds numbers from 600 to 2300.

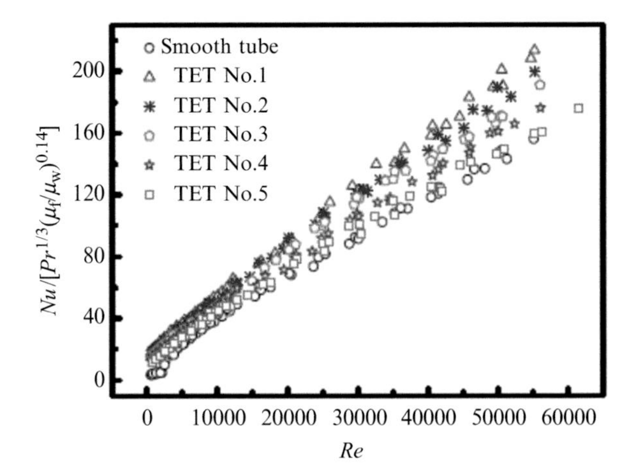

Fig. 3.18 Variation of Nusselt number with Reynolds number for various tubes

Fig. 3.19 Variation of Nusselt number ratio (Nu_e/Nu_s) with Reynolds number for the TETs

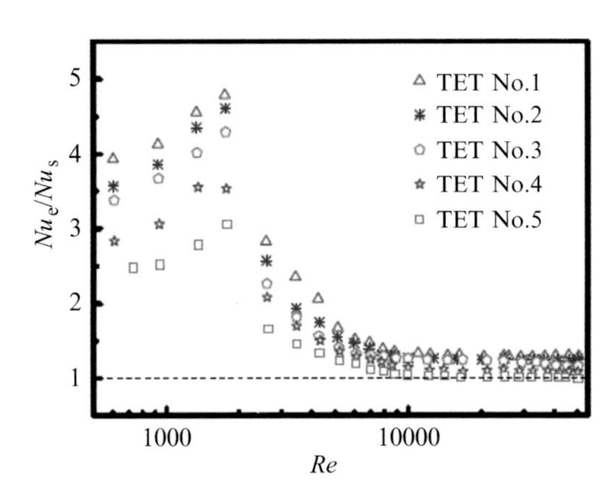

3.3.2 *Frictional Loss Aspect of Twisted Tubes*

Variation of friction factor with Reynolds number is presented in Fig. 3.20 as the index of hydraulic performance. It is found that the friction factors for all the TETs are greater than those for the smooth tube. It is further noticed that friction factor depicts hyperbolic variation with the Reynolds number i.e. friction factor values are high, say up to Reynolds number 15000, then onwards it decreases. TET No. 3 has demonstrated the greatest friction factors since it has the highest aspect ratio. In this experiment it is found that the influence of aspect ratio on the flow resistance is more than the twist pitch.

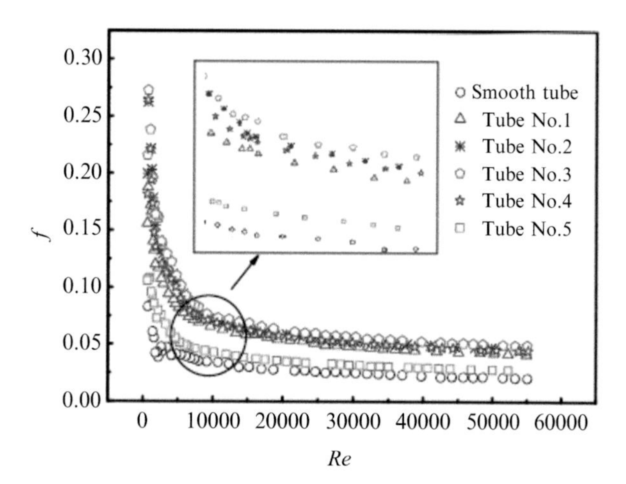

Fig. 3.20 Variation of friction factor with Reynolds number for the TETs

Chapter 4
Advances in Compound Techniques (Fourth Generation Heat Transfer Technology)

Abstract The advances in heat transfer enhancement techniques in all its aspects have been dealt with.

Keywords Heat transfer enhancement techniques • Active technique • Passive technique • Combined techniques • Heat transfer coefficient and friction factor correlations • Performance evaluation criteria

Researchers are now exploring the possibilities of combining attributes of different enhancement techniques with the hope to get more heat transfer than the individual technique will do. In this context experimental works related to compound techniques of heat transfer enhancement have been discussed here.

4.1 Helical-Ribbed Tube with Double Twisted Tape Inserts

In this approach heat transfer characteristics of ribbed tube fitted with the twin twisted tape insert are compared with those from the smooth tube and the ribbed tube acting alone. This combination of the modified tube and twisted tapes has been investigated with the prospect of high heat transfer enhancement. The work has been conducted in the turbulent flow regime, Re from 6000 to 60,000 using water as the test fluid. The details of twin twisted tape and ribbed tube has been shown in the Fig. 4.1.

It is to be noted that when the ribbed tube is arranged in similar directions of the helical swirl of the twisted tape and the helical rib motion, the tube is called co-swirl. Effects of the co-swirl motion of the ribbed tube and the double twisted tapes with various twist ratios on heat transfer and friction characteristics have been examined.

S.K. Saha et al., *Advances in Heat Transfer Enhancement*,
DOI 10.1007/978-3-319-29480-3_4

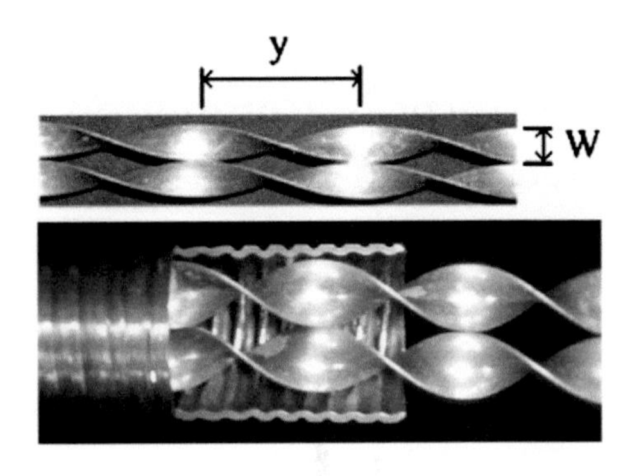

Fig. 4.1 Double twisted tapes and helically-ribbed tube with double twisted tape insert

4.1.1 Heat Transfer Performance of Ribbed Tube with Twin Twisted Tape

Nusselt number (Nu) variation with Reynolds number has been displayed (Fig. 4.2a) as the heat transfer performance indicator for ribbed tube with twin twisted tape, the performance is compared with ribbed tube and plain tube and is reflected in the form of Nu/Nu_0 ratio variation against Reynolds number (Fig. 4.2b).

It can be observed that Nu increases with the increase in Re while the Nu/Nu_0 shows an opposite trend for all cases. On the other hand, combined technique i.e. ribbed tube having twin twisted tape shows improved heat transfer rate results than the ribbed tube acting alone or the smooth tube. This improved performance may be attributed to the more effective stagnant layer disruption capability of twin twisted tape assisted by the ribbed wall roughness, compared to those caused by ribbed tube or plain tube individually. Heat transfer augmentation using combined technique is found to improve by 4–75 % with respect to the ribbed tube. Of course, this variation is due to different combination of Re and twist ratio value. On the other hand, the augmentation has been observed 150–320 % more compared to the plain tube.

4.1.2 Effect on Friction Factor

Figure 4.3a, b show the variation of friction factor (f) with Reynolds number and comparison of friction factor (f/f_0) among ribbed tube with twin twisted tape insert, only ribbed tube and the plain tube as Reynolds number is varied. It is observed that for all cases, friction factor (f) shows a decreasing trend. Of course, frictional losses for ribbed tube with twin twisted tape insert and with only ribbed tube are higher than the plain tubes (the f of the ribbed tube alone is around 77–207 % above that of the smooth tube while the f of the inserted ribbed-tube is found to be approximately 1.7–3.6 times the ribbed tube alone and to be about 6.0–19.2 times the smooth tube, depending on the Re and y values.).

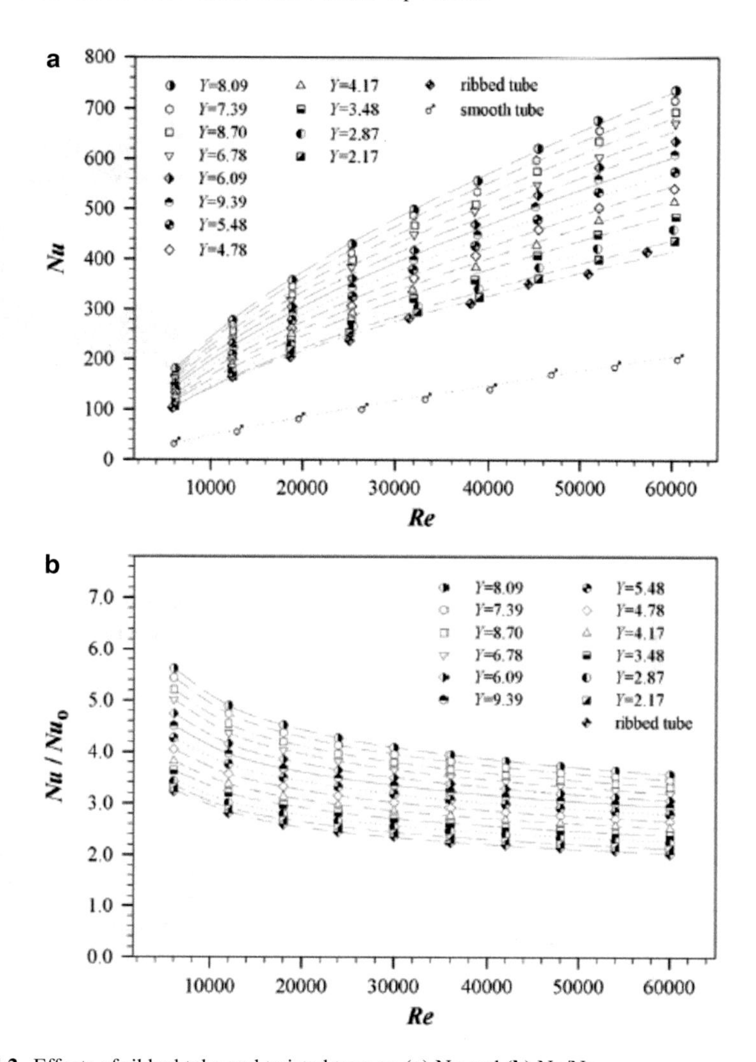

Fig. 4.2 Effects of ribbed tube and twisted tape on (**a**) Nu and (**b**) Nu/Nu_0

4.1.3 Performance Evaluation of Ribbed Tube with Twin Twisted Tape

In order to assess the practical use of the enhanced tube, the performance of the enhanced tube is evaluated relative to the smooth tube at an identical pumping power in the form of thermal performance enhancement factor (TEF) is evaluated using the equation given by

$$\text{Thermal enhancement factor}\left(\text{TEF}\right)=\left(\frac{Nu}{Nu_o}\right)/\left(\frac{f}{f_o}\right)^{1/3}$$

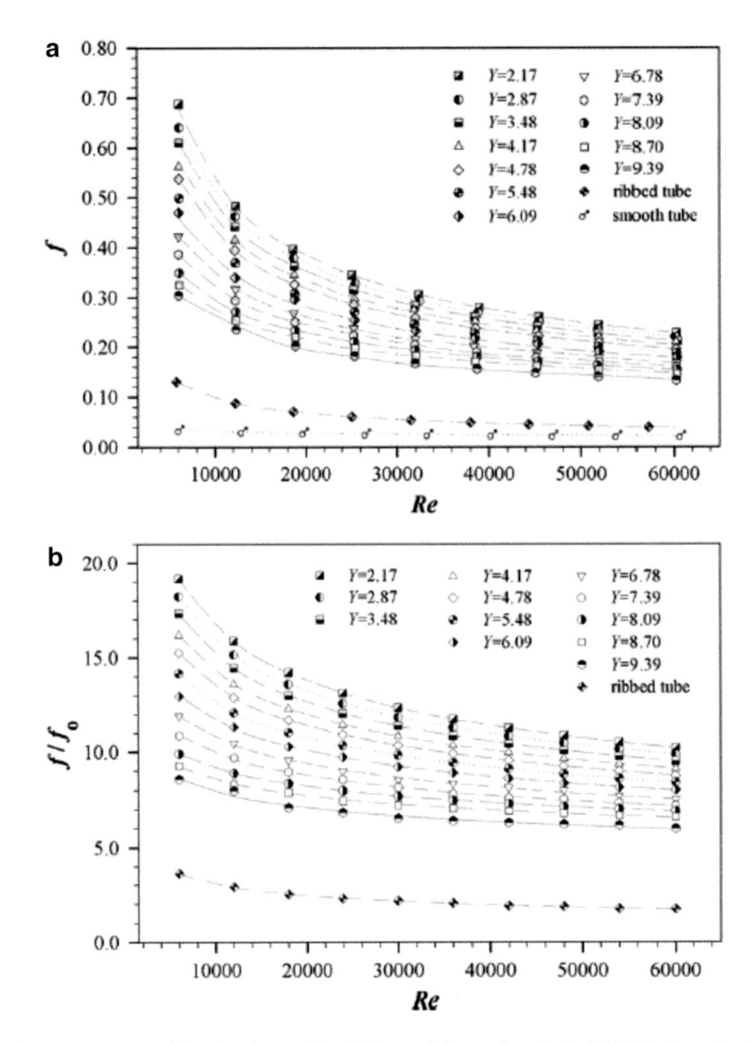

Fig. 4.3 (**a**) Variation of friction factor (f) with Reynolds number (Re). (**b**) Variation of ratio (f/f_0) with Reynolds number (Re)

Figure 4.4 shows that the thermal enhancement factor is exhibiting a decreasing trend with increase in Re. Its value is found to be about 2.6 and 1.9 at the lowest and highest Re values, respectively. On the other hand, depending on Re, the TEF of the ribbed tube alone is around 1.6–2.2, depending on Re. Hence it may be concluded that from energy saving point of view the compound technique of ribbed tube with twin twisted tape inserts is useful when operated at low Reynolds number.

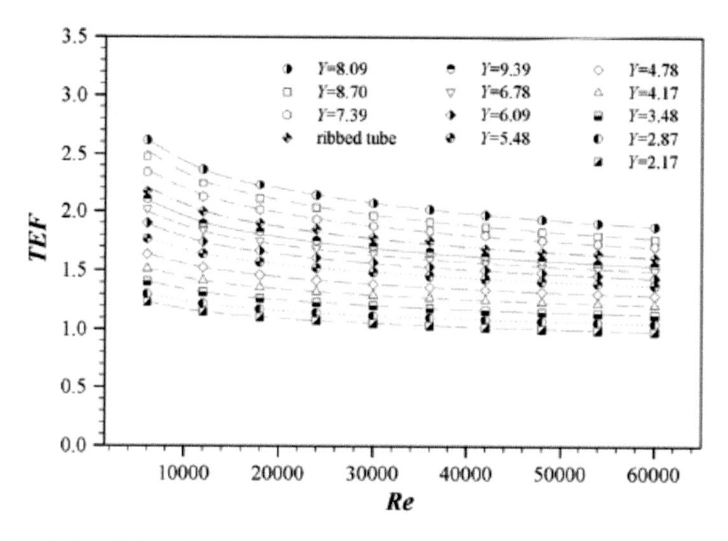

Fig. 4.4 Variation of thermal enhancement factor (TEF) with Re

4.1.4 Empirical Correlations

Following empirical correlations have been developed for co-swirl ribbed-tube with double twisted tapes

$$\mathrm{Nu} = 0.238\,\mathrm{Re}^{0.627}\,\mathrm{Pr}^{0.3}\,\mathrm{Y}^{0.346}$$

$$\mathrm{f} = 31.675\,\mathrm{Re}^{-0.4}\,\mathrm{Y}^{-0.458}$$

4.1.5 Influence of Combined Non-uniform Wire Coil and Twisted Tape Inserts

Previous research work on wire coil and twisted tape inserts reveals that twisted tape insert disturbs the entire flow field while the wire coil insert mainly produces disturbance in the flow near the wall. Obviously, from heat transfer point of view, twisted tape inserts are more promising but if we compare pumping power penalty, wire coil will be a better option. Hence in the situation where pressure drop is a crucial restriction, wire coil becomes more useful due to its less pressure drop penalty. Scientist and Researchers have explored the possibility of augmenting heat transfer by combining wire coil with twisted tape inserts. Experimental investigations have been carried to observe the influence of combined non-uniform wire coil and twisted tape inserts on thermal performance characteristics. The experiments were conducted in a turbulent flow regime with Reynolds numbers ranging from 4600 to 20,000 using air as the test fluid. Steel wire coil (dia. 4.8 mm) having

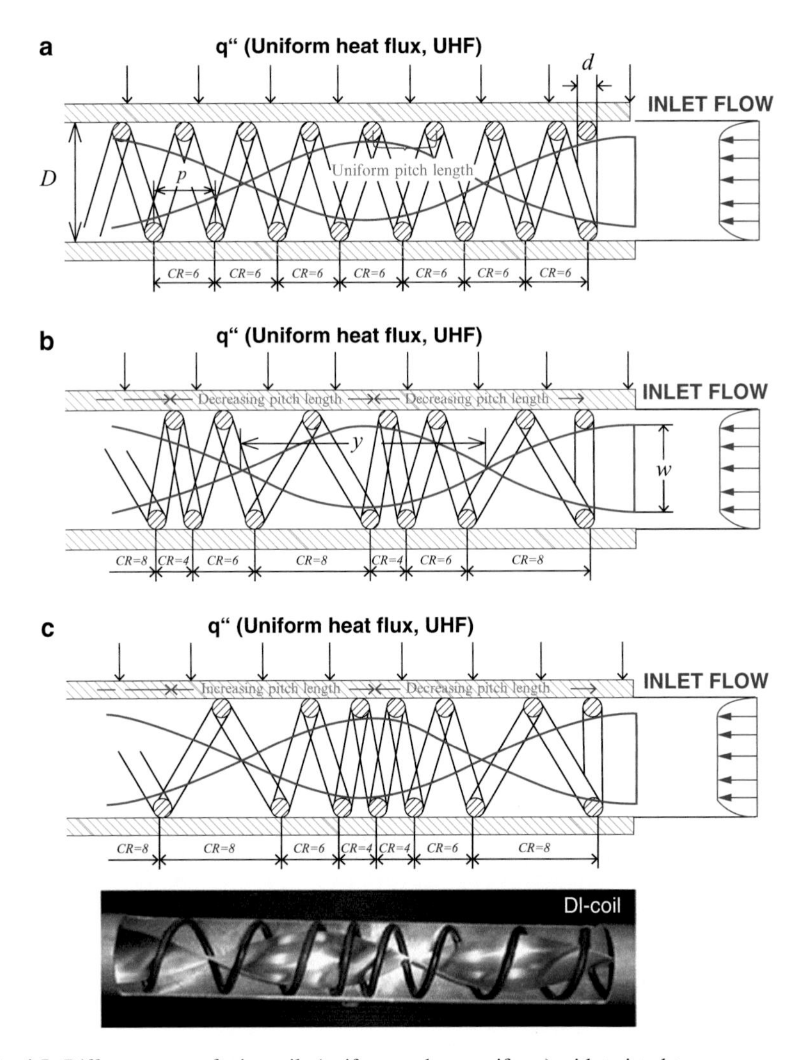

Fig. 4.5 Different types of wire coils (uniform and non uniform) with twisted tape

varying coil pitch ratio (p/d) of two forms have been used along with Aluminum twisted tapes with twist ratios of 3 and 4. The details of wire coils and twisted tapes are shown in Fig. 4.5.

4.1.6 Heat Transfer Augmentation and Comparison

Figure 4.6 shows the variation of dimensionless heat transfer coefficient (Nu) with Reynolds number (Re) for the combination of twisted tape (TT) and non-uniform wire coil (the wire coil with varying three coil pitch ratio, D-coil/DI-coil) and

Fig. 4.6 Variation of Nu with Re

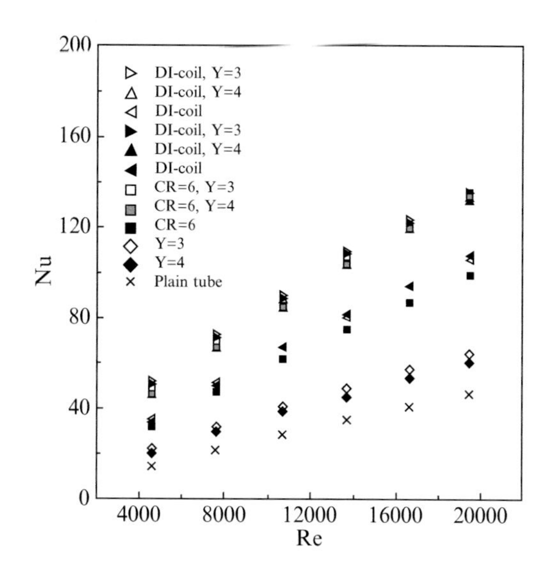

uniform wire coil (the wire coil with constant coil pitch ratio). Nu is found to have increasing trend with increasing Re for all combinations and at any particular Reynolds number, the Nusselt numbers for the tubes equipped with compound enhancement devices (D/DI-coil with TT) are higher than those with each device alone and also the plain tube.

4.1.7 Dimpled Tube with Twisted Tape Inserts

Experimental investigations using water as working fluid have been carried out to find thermal and friction characteristics when dimpled tube is coupled with twisted tape. The effects of pitch and twist ratios on the heat transfer coefficient and pressure loss characteristics in the fully developed turbulent flow (12,000 <Re< 44,000) of a dimpled tube with a twisted tape insert have been examined. During the experiment, the dimple diameter (d) and dimple depth (e) were kept at a constant value of 3 and 2 mm, respectively. Figure 4.7 shows the diagram of concentric tube heat exchanger and the geometry of dimpled tube combined with a twisted tape.

Two different pitch ratios (PR = p/D = 0.7 and 1.0) of dimpled surface were used while three twisted tapes with three different twist ratios (y/w = 3, 5, and 7) were used. The performance of heat exchanger with dimpled tube with twisted tape is presented in Fig. 4.8a and b and in this plot, performance of proposed combination has also been compared with dimpled tube and with the plain tube when they are acting alone. The plot reveals that Nusselt number (Nu) and friction factor (f) characteristics in dimpled tubes combined with a twisted tape (y/w = 3), increases with increasing Reynolds number and the Nusselt number of the dimpled tube with the twisted tape insert is 15–56 % higher than that in the dimpled tube acting alone and 66–303 % higher than that in the plain tube.

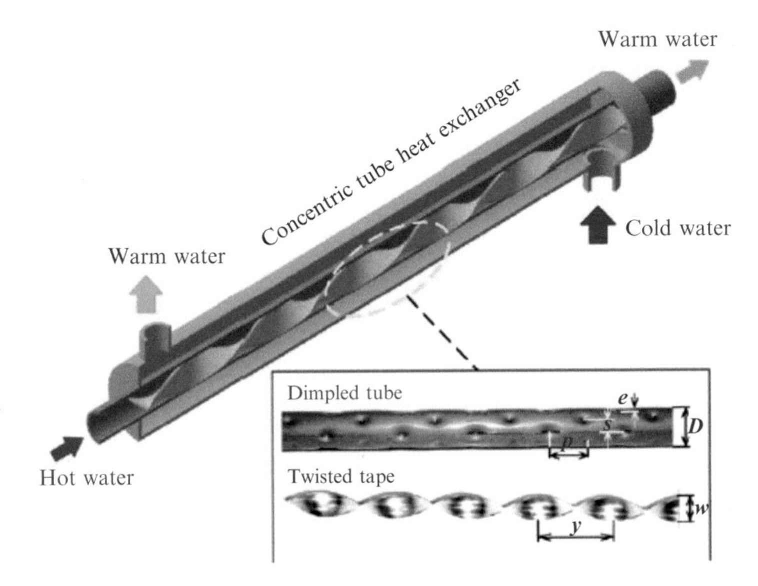

Fig. 4.7 Dimpled tube and twisted tape

The augmentation may be attributed to the combined effect of induced swirl flow due to twisted tape and reduction in boundary layer thickness due to wall roughness (dimpled surface). But the friction factor, on the other hand, in the combined devices is up to 2.12 times of that in the dimple tube acting alone and 5.58 times of that in the plain tube. Further, from same plot the effect of dimple pitch on heat transfer may be noticed and it is found that the Nusselt number in the dimpled tube combined with twisted tape for pitch ratio (PR) = 0.7 is considerably higher than those for higher pitch ratio (PR = 1.0). This is so, since when the dimple pitch reduces, wall roughness increases, resulting in more turbulence in the boundary layer hence improving the heat transfer while increasing the frictional losses.

4.1.8 *Influence of Twist Ratio on Heat Transfer and Frictional Losses*

It was found that the Nusselt number of a dimpled tube when combined with a twisted tape insert, attained highest value for the lowest value of twist ratio (among y/w = 3, 5 and 7), as shown in Fig. 4.9

This improvement in heat transfer occurs, because as the twist ratio decreases, the fluid retention period increases, also turbulence intensifies with reduced twist ratio. The friction factor, on other hand, shows an increasing trend as the twist ratio reduces (Fig. 4.10). Approximately, the friction factors for employing the dimpled tube combined with a twisted tape at the smallest twist ratio (y/w = 3) are found to be 9.7 and 18.3 % over those at the twist ratio, y/w = 5 and 7, respectively.

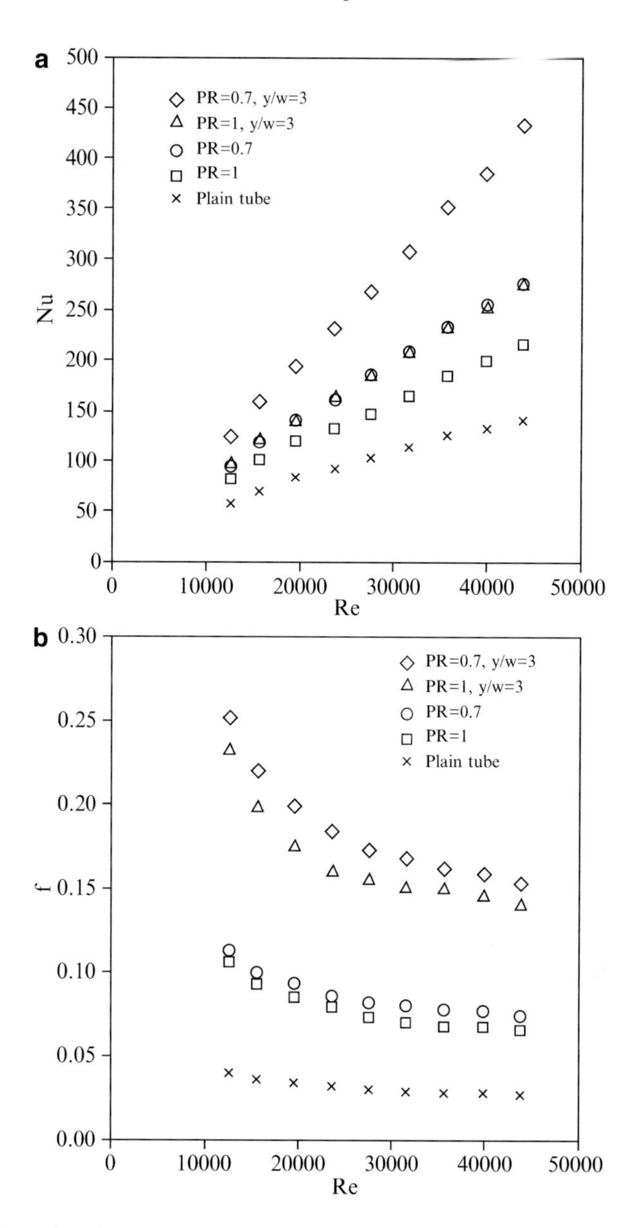

Fig. 4.8 Effects of combining dimpled tube with twisted tape on (**a**) Nusselt number and (**b**) friction factor

Following empirical correlations, Nusselt number and friction factor have been proposed for a dimpled tube combined with a twisted tape.

$$Nu = 0.014\,Re^{0.9}\left(PR\right)^{-0.93}\left(y/w\right)^{-0.12}Pr^{0.3}$$

$$f = 9.1\,Re^{-0.37}\left(PR\right)^{-0.11}\left(y/w\right)^{-0.2}$$

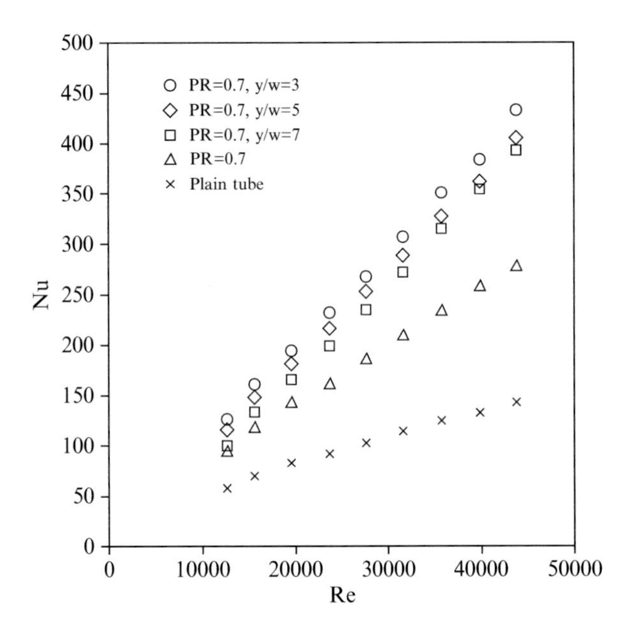

Fig. 4.9 Effect of twist ratio on heat transfer enhancement for PR = 0.7

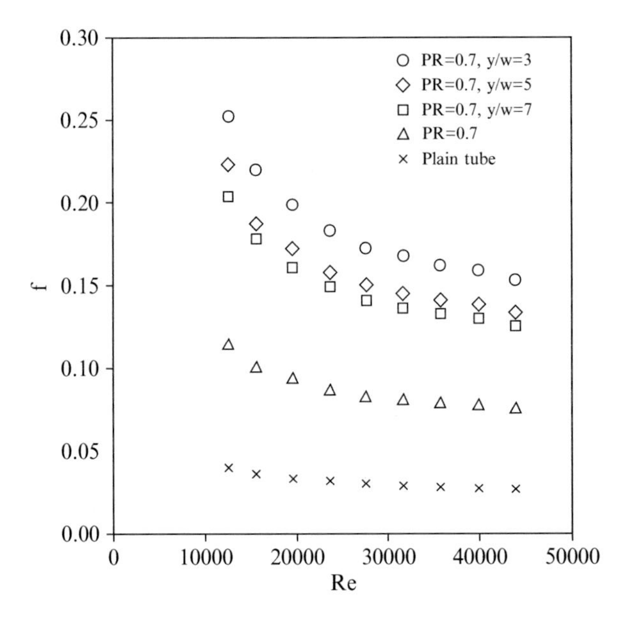

Fig. 4.10 Effect of twist ratio on friction factor for PR = 0.7

4.1.9 *Integral Type Wall Roughness with Wavy Strip Inserts*

The performance of centre-cleared wavy strip inserts in laminar flow in combination with integral helical rib roughness, using servotherm (medium) oil as working fluid has been investigated experimentally. Figures 4.11 and 4.12 show the geometries of circular tube having integral type wall roughness and wavy strip insert.

The effect of roughness geometry (i.e. rib height, pitch and angle) and centre clearance on heat transfer and frictional losses have been reported based on the experimental data that have been generated for the following values of the parameters:

Dimensionless centre clearance on the wavy strip, (C/D) = 0, 0.2, 0.4, 0.6.
Rib helix angle (θ) = 30°, 60°.
Dimensionless rib pitch, (P/e) = 35.5422, 20.0869.
Rib dimensionless height (e/D) = 0.07581, 0.1052.

Figures 4.13 and 4.14 show the effect of centre clearance size on heat transfer and frictional loss and it is found that, initially Nusselt number and friction factor decreases with increase in dimensionless centre clearance (c) up to value 0.4, and then onwards, however the influence of centre clearance on the Nusselt number and friction factor is not significant.

The decrease in friction factor is due to increase in free flow passage with increase in centre clearance while reduction in Nusselt number is due to reduced turbulence as centre clearance increases. Further, the effect of rib height and pitch is demonstrated in Figs. 4.15 and 4.16.

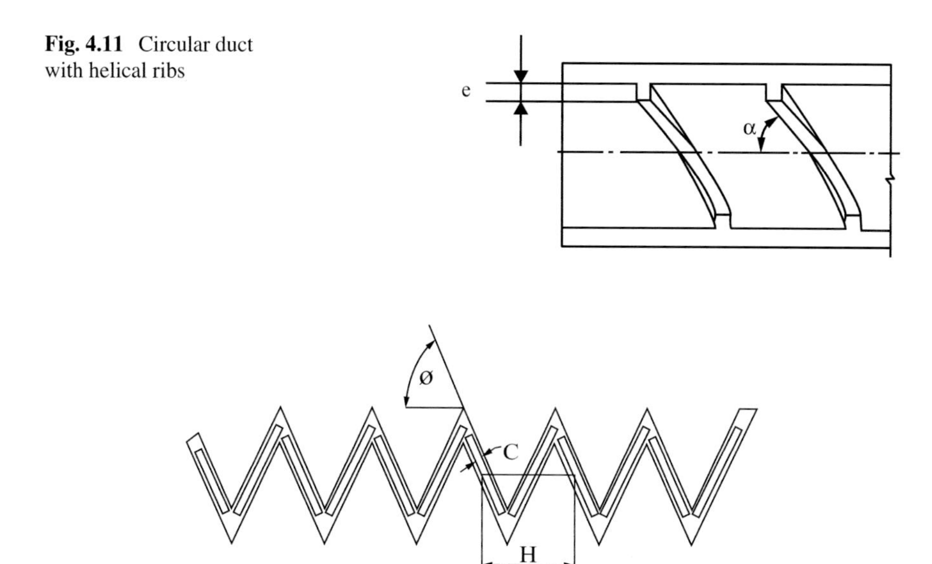

Fig. 4.11 Circular duct with helical ribs

Fig. 4.12 Centre-cleared wavy strip

Fig. 4.13 Effect of centre clearance of wavy strip insert on Nusselt number

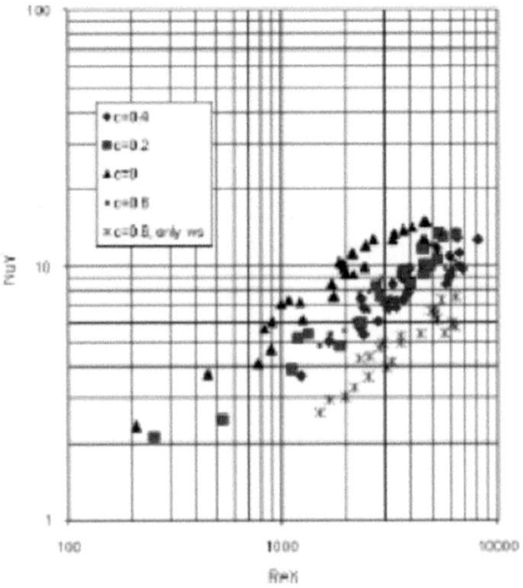

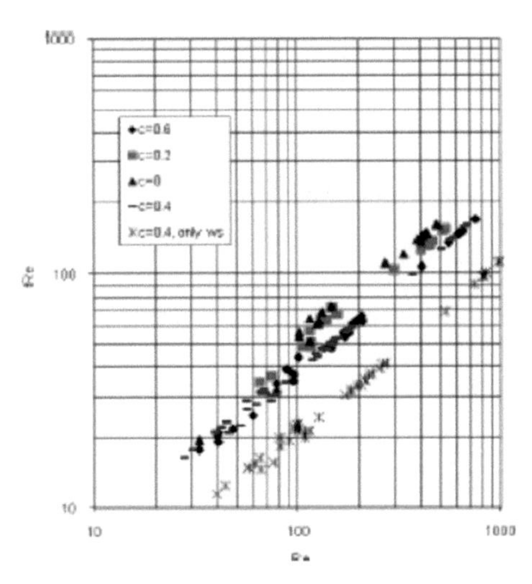

Fig. 4.14 Effect of centre clearance of wavy strips on friction factor

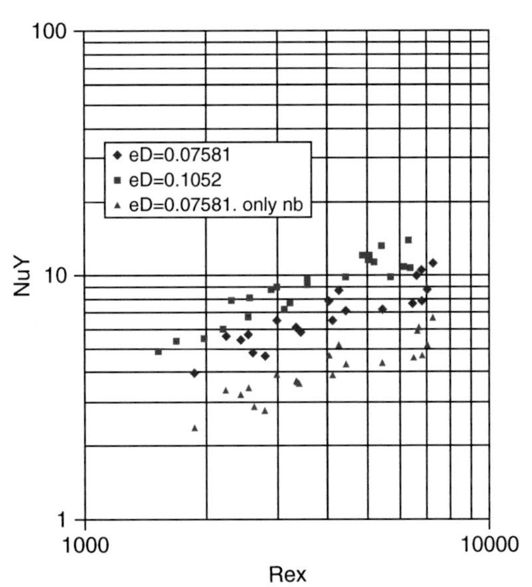

Fig. 4.15 Effect of rib height on Nusselt number

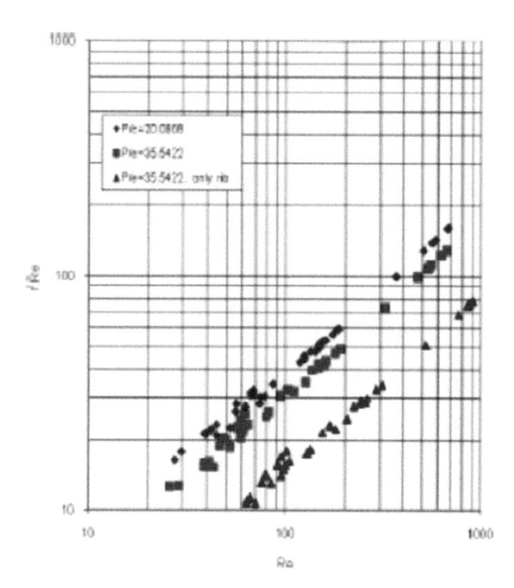

Fig. 4.16 Effect of rib pitch on friction factor

It was observed that both Nusselt number and friction factor increase with increase in rib height because flow passage blockage increases with increase in rib height and this results in increased flow velocity and turbulence, causing improved heat transfer coefficient accompanied with increased frictional loss. On the other hand, as the pitch of the rib is reduces, the heat transfer rate improves.

Chapter 5
Nano-Fluids, Next-Generation Heat Transfer

Abstract The advances in heat transfer enhancement techniques in all its aspects have been dealt with.

Keywords Heat transfer enhancement techniques • Active technique • Passive technique • Combined techniques • Heat transfer coefficient and friction factor correlations • Performance evaluation criteria

In passive techniques, heat transfer augmentation is achieved by following different methods like flow passage surface modification, adding roughness to increase surface area and turbulence. However, poor thermal conductivity of heat transfer fluid has always remained the weaker point. In an effort to overcome this, knowing the fact that solids possess higher thermal conductivity, idea of adding certain solid particles into the base liquid was framed. In the wake of this, when this innovative idea was executed by using suspensions of millimeter or micro sized particles in conventional heat transfer fluids, problem such as poor suspension stability leading to channel clogging was noticed. As a solution to this problem, the size of the additives is further reduced to nanometer sized particles. The fluids with these solid-particle suspended in them are called 'nano-fluids'. Nano-fluids are solid-liquid composite materials consisting of solid nanoparticles or nanofibers with sizes typically of 1–100 nm suspended in liquid. The suspended metallic or nonmetallic nanoparticles change the transport properties and heat transfer characteristics of the base fluid. With the advent of nanotechnology, thermal sciences researchers find an alternative to the conventional fluids used in heat exchange activities. Nano-fluids are suspensions of nanoparticles in fluids that show significant enhancement of their properties at modest nanoparticle concentrations (Fig. 5.1).

5.1 Classification of Nano-Fluids System

It is possible to quote many examples present in the nature itself in the form of nanoparticles, some of which are shown in the Fig. 5.2 with length scale.

S.K. Saha et al., *Advances in Heat Transfer Enhancement*,
DOI 10.1007/978-3-319-29480-3_5

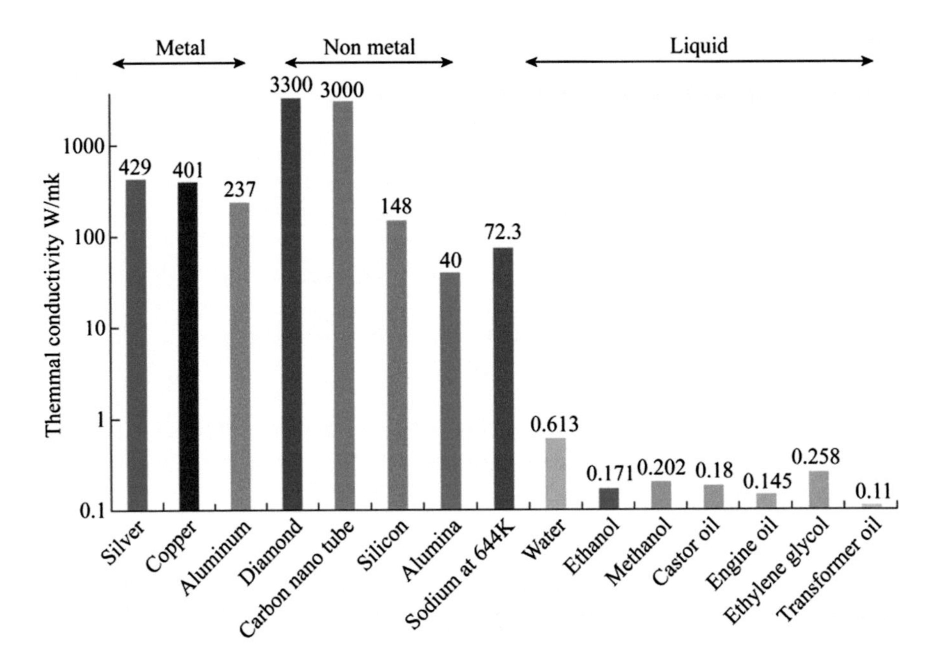

Fig. 5.1 Comparison of thermal conductivity for different materials

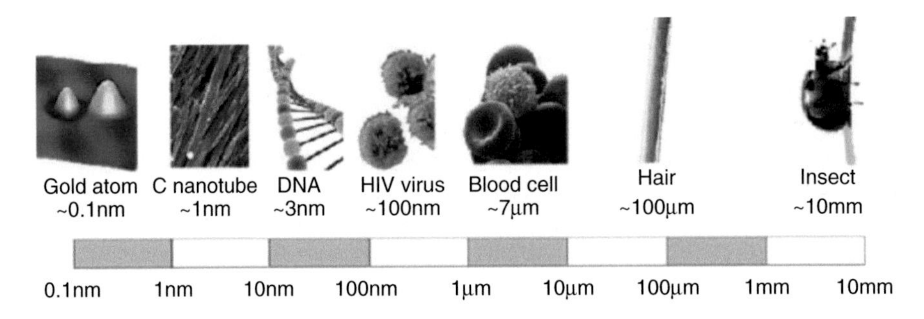

Fig. 5.2 Length scale and some examples of nanoparticles

Nevertheless, depending upon the type of nanoparticle material, nano-fluids may be addressed as (1) ceramic nano-fluids (2) metallic nano-fluids (3) carbonaceous nano-fluids and (4) nanoemulsions.

Among above stated class of nano-fluids, most of the investigations are related to ceramic nano-fluids because of their wide availability at low cost and chemical stability.

5.2 Distinct Features of Nano-Fluids

Nano-fluids have an unprecedented combination of the four characteristic features desired in energy systems (fluid and thermal systems):

1. Increased thermal conductivity at low nanoparticle concentrations
2. Strong temperature-dependent thermal conductivity
3. Non-linear increase in thermal conductivity with nanoparticle concentration
4. Increase in boiling critical heat flux (CHF)

These characteristic features of nano-fluids make them suitable for the next generation of flow and heat-transfer fluids.

5.3 Preparation Methods for Nano-Fluids

For producing high quality nano-fluids, some special attributes are prerequisites, e.g., even and stable suspension, durable suspension, negligible agglomeration of particles, no chemical change of the fluid, etc. Fundamentally there are two approach (1) Two-step process (2) one step process

The **two-step method** is extensively used in the synthesis of nano-fluids considering the available commercial nanopowder supplied by several companies. In this method, nanoparticles are first produced and then dispersed in the base fluids. Generally, ultrasonic equipment is used to intensively disperse the particles and reduce the agglomeration of particles. Making nano-fluids using the two-step processes is challenging because individual particles tend to quickly agglomerate before complete dispersion can be achieved. This agglomeration is due to attractive Vander Waals forces between nanoparticles, and the agglomerations of particles tend to quickly settle out of liquids. In fact, agglomeration is a critical issue in all nanopowder technology, including nano-fluids technology, and a key step to success in achieving high-performing heat transfer nano-fluids is to produce and suspend nearly monodispersed or non-agglomerated nanoparticles in liquids.

The **one-step method** simultaneously produces and disperses nanoparticles into the base fluids, while the two-step method disperses previously manufactured nanoparticles in a base fluid. No matter which one of these two methods is employed, the production of nanoparticles inherently involves reduction reactions or ion exchange. Ions and other reactions products are then dispersed in the fluid together with the nanoparticles, since they are almost impossible to separate from their surroundings. For nano-fluids containing high-conductivity metals such as copper, a one-step technique is preferred to the two-step process to prevent oxidation of the particles. An advantage of the one step technique is that nanoparticle agglomeration is minimized, while the disadvantage is that only low vapor pressure fluids are compatible with such a process. Although the one-step physical methods have produced nano-fluids in small quantities for research purposes, they are unlikely to become the mainstay of commercial nano-fluids production. They would be difficult to scale up for two reasons. Processes that require a vacuum significantly slow the production of nanoparticles and nano-fluids, thus limiting the rate of production. Furthermore, producing nano-fluids by these one step physical processes is expensive.

While most nano-fluids productions to date have used one of the above-described (one-step or two-step) techniques, other techniques are available depending on the particular combination of nanoparticle material and fluid. For example, nanoparticles with specific geometries, densities, porosities, charge, and surface chemistries can be fabricated by templating, electrolytic metal deposition, layer-by-layer assembly, micro-droplet drying, and other colloid chemistry techniques.

5.4 Application of Nano-Fluids in Automobile Heat Exchangers as Coolant

Two requirements of automobile heat exchangers (called radiators) are critical: (1) coolant must be able to quickly remove heat from the engine components (2) radiator must be light weighted and should occupy minimum space in engine cabin. In automobiles, the cooling system is responsible for thermal management of the engine block and passenger compartments. The engine life and its performance depend on coolant temperature. Usually a mixture of ethylene glycol and water is used as a coolant in the radiator of automobile engines. These fluids possess poor heat transfer performance compared to water because of lower thermal conductivity. It is observed that conventional fluids are unable to meet the increasing demand in cooling of automobile engines but with the advent of nanotechnology, it seems possible to prepare coolants with improved thermal properties by addition of small particles with higher thermal conductivity. The fluids containing a suspension of nanometer sized particles found to possess substantially higher thermal conductivity compared to their base fluids. The use of nano-fluids as coolants would allow for smaller size and better positioning of the radiators. Owing to the fact that there would be less fluid due to the higher efficiency, coolant pumps could be shrunk and engines could be operated at higher temperatures allowing for more horsepower while still meeting stringent emission norms. Several researchers have studied the influence of eight parameters, namely particle volume concentration, particle material, particle size, particle shape, base fluid material, temperature, additive, and acidity on nano-fluids thermal conductivity enhancement.

The application of Al_2O_3/water nano-fluids in the car radiator has been tested and the tube-side heat transfer coefficient has been calculated. Interestingly, an enhancement of 45 % compared to pure water application under highly turbulent flow condition has been observed. During the experiment, automobile radiator of louvered fin-and-tube type having fins and tubes made of aluminum has been used; the results are shown in the Fig. 5.3.

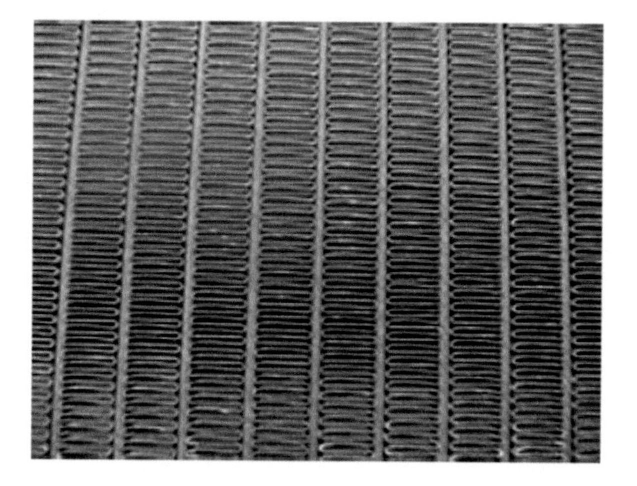

Fig. 5.3 Automobile radiator of louvered fin-and-tube type

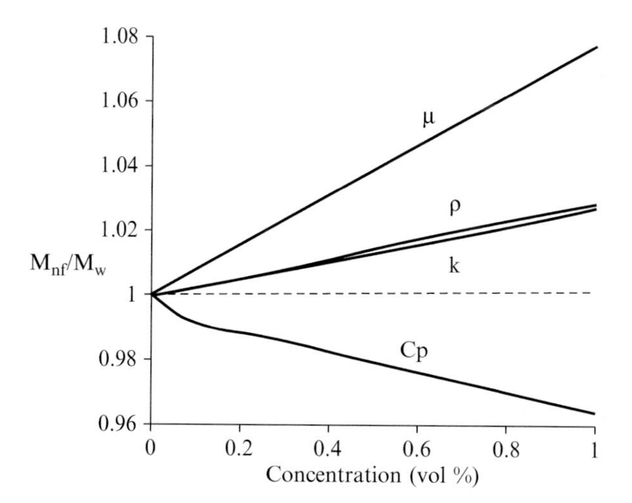

Fig. 5.4 Dimensionless physical properties of nano-fluids in comparison to those of pure water

The influence of Alumina nanoparticle concentration on the physical properties of nano-fluids has been studied by plotting the variation of the ratios of physical properties of the nano-fluid to those of pure water as a function of nano-particle concentration. Figure 5.4 shows strong influence of nano-particle concentration on physical properties of nano-fluids. Density and thermal conductivity has increased while specific heat has decreased slightly in comparison to base fluid. An undesirable influence from the heat transfer point of view on viscosity is observed i.e. significant increase in viscosity is noticed.

5.4.1 *Effect of Augmentation of Nano-Particle Concentration on Radiator Cooling Performance*

Keeping fluid inlet temperature at 44 °C and varying the nano-particle volume concentration, as shown in Fig. 5.5, it is found that fluid outlet temperature has decreased with increase in nano-particle volume concentration i.e. thermal performance of cooling system has improved (since from a practical viewpoint for every cooling system, at equal mass flow rate the more reduction in working fluid temperature indicates a better thermal performance of the cooling system).

Further, replacing water with nano-fluids and its subsequent effect on heat transfer enhancement is reflected in Fig. 5.6. Following are the observations:

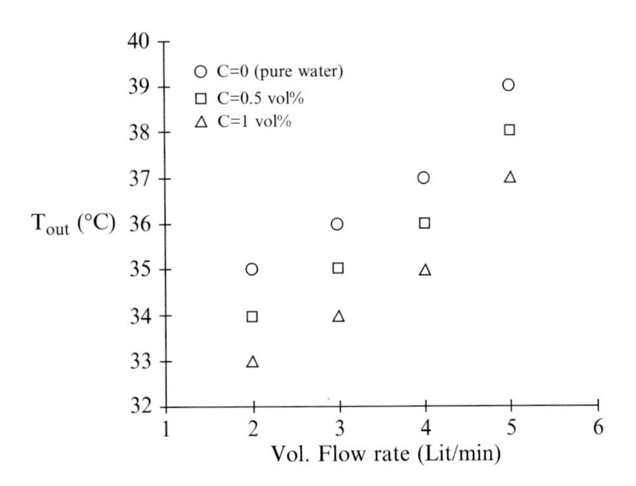

Fig. 5.5 Comparison of the radiator cooling performance when using nanofluid (0.5 and 1 vol.%) and pure water

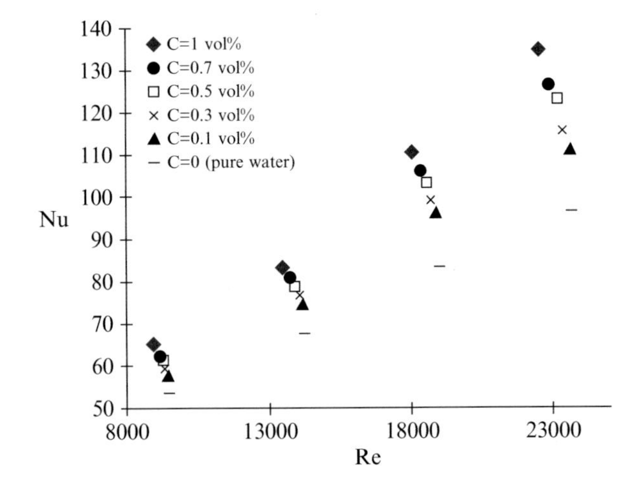

Fig. 5.6 Nu number variations of nano-fluids at different concentrations as a function of Re number (T_{in}=44 °C)

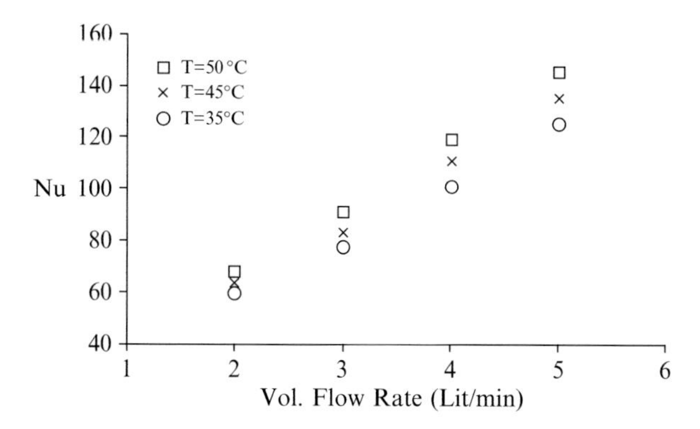

Fig. 5.7 Effect of nano-fluid inlet temperature on the Nu numbers for the concentration of 1 vol.%

1. Addition of nanoparticles, irrespective of its concentration, results in increase in Nusselt number with increase in Reynolds number.
2. Concentration of nano-particle has an important bearing on the heat transfer efficiency. As the concentration becomes greater, heat transfer coefficient becomes larger and it is found that by the addition of only 1 vol.% of Al_2O_3 nano-particle into the pure water, an increase of about 30–45 % in comparison with the pure water heat transfer coefficient has been recorded.

5.4.2 *Effect of Fluid Inlet Temperature on Heat Transfer Performance of the Automobile Radiator*

It is clear from Fig. 5.7 that an increase in the fluid inlet temperature (in the range of 35–50 °C) at a fixed concentration of nanoparticles (1 vol.%), slightly improves the heat transfer coefficient. This small variation in Nu may be attributed to the effect of temperature on the physical properties and also to the increased effect of radiation.

Chapter 6
Effect of Ultrasounds on Thermal Exchange

Abstract The advances in heat transfer enhancement techniques in all its aspects have been dealt with.

Keywords Heat transfer enhancement techniques • Active technique • Passive technique • Combined techniques • Heat transfer coefficient and friction factor correlations • Performance evaluation criteria

Sound waves having frequencies higher than those can be heard by human (i.e. above 16 kHz) is called ultrasound and depending upon frequency or power, sound waves may be classified as (1) low frequency ultrasound (between 20 and about 100 kHz) or high power ultrasound and (2) high frequency ultrasound (above 1 MHz) or low power ultrasound. The high power ultrasound is capable of modifying the medium in which it is propagated. Propagation of high power ultrasound produces cavitations, micro-streaming, heating and surface instability effects at liquid-liquid, liquid-gas and liquid-solid interfaces. These attributes are very attractive from heat exchangers performance point of view since two fold advantages i.e. heat transfer augmentation and reduction in heat exchanger fouling can be achieved by propagating high power ultrasound waves. Therefore, high power ultrasound finds uses in various processes like cleaning, plastic welding, sonochemistry etc. The low power ultrasound, on other hand, does not affect the medium of propagation. Consequently, it is especially used for medical diagnosis or nondestructive material control. Various applications of ultrasound according to their frequency or power are shown in the Fig. 6.1.

6.1 Understanding Enhancement Mechanism

The ultrasonic system transforms electrical power into vibrational energy, i.e. mechanical energy. The mechanical energy is then transmitted into a sonicated medium. A part of input energy is lost through conversion into heat, and another part produces cavitations. A fraction of cavitational energy produces chemical, physical, or biological effects. The conversion of electrical energy into ultra-sonic effects is shown in the Fig. 6.2.

S.K. Saha et al., *Advances in Heat Transfer Enhancement*,
DOI 10.1007/978-3-319-29480-3_6

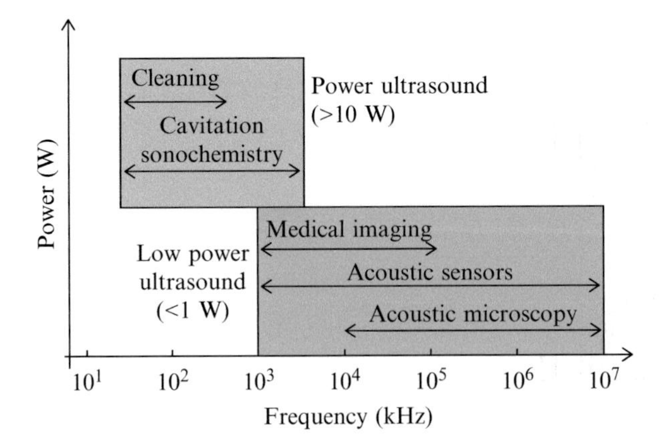

Fig. 6.1 Utilizations of ultrasound according to frequency and power

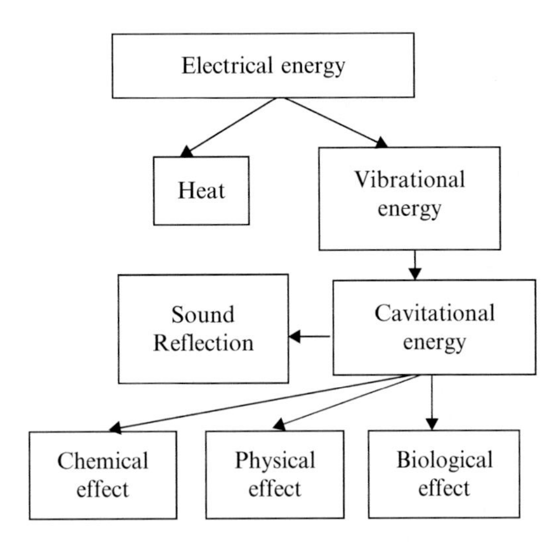

Fig. 6.2 Conversion of electrical energy into physical, chemical and biological effects

From heat transfer point of view, the relevant effects of ultrasound, when propagated in liquid medium is shown in the Fig. 6.3. Out of four stated effects, two of them, of major importance for heat transfer enhancement, are acoustic cavitations and acoustic streaming. When the high power ultrasound waves are propagated through the liquid medium, local pressure decreases sufficiently below the vapor pressure during the rarefaction period of the sound wave, resulting in formation of bubbles and as these bubbles move to high pressure region they collapse, producing intense local heating and high pressure. The phenomenon of formation, growth and collapse is termed as acoustic cavitations and heat transfer enhancement due to this phenomenon is self-explanatory in Figs. 6.3 and 6.4. A bubble collapse near a solid-liquid interface disrupts thermal and velocity boundary layers, reducing thermal resistance and creating micro-turbulence. This intensifies heat transfer.

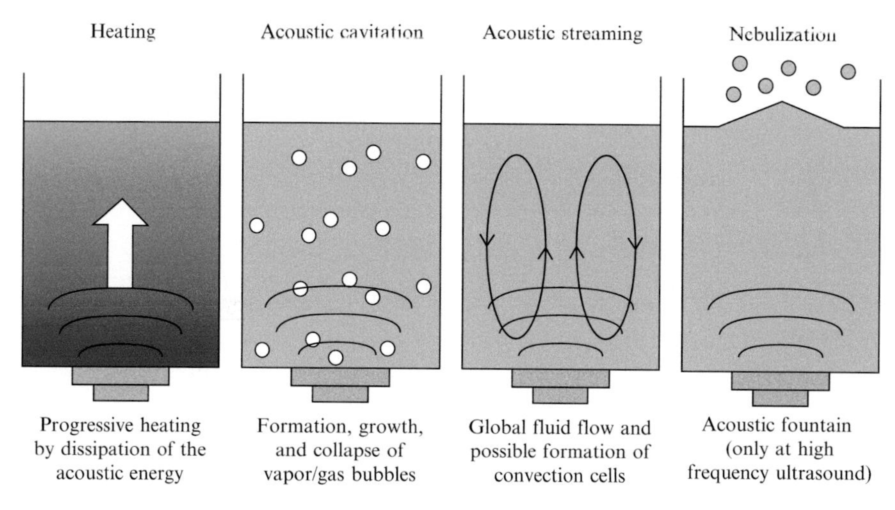

Fig. 6.3 Four effects resulting from ultrasound propagation in a liquid

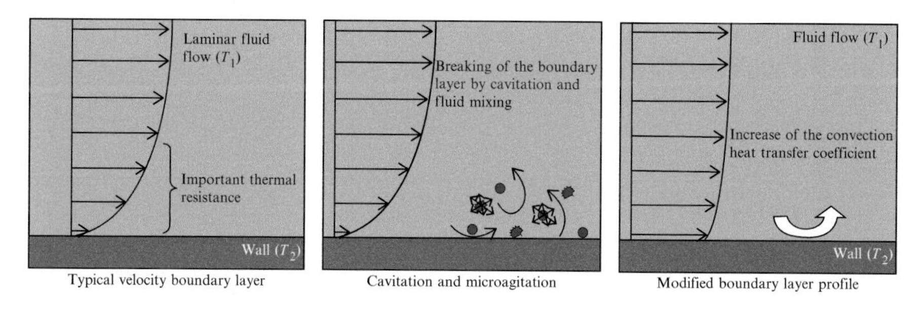

Fig. 6.4 Mechanism of heat transfer enhancement by acoustic cavitation

Acoustic streaming phenomenon occurs when a micro bubble, surrounded by a liquid medium, undergoes direct oscillatory action when exposed to ultrasound. Oscillatory action causes rapid, toroidal eddy currents form as a result of the displacement of liquid around the bubble. These eddy currents decrease in size as the ultrasonic frequency is increased; that is, less micro streaming occurs at higher frequencies. Micro-streaming, therefore, is a consequence of low-frequency, low-pressure response to oscillations of micro bubbles.

A new kind of ultrasonically-assisted heat exchanger as shown in Fig. 6.5 has been developed. Ultrasound can be used efficiently as a heat transfer enhancement technique, even in such complex systems as heat exchangers. The heat exchanger used is a double-pipe configuration, i.e. it is made of two concentric straight pipes inserted one into the other as shown in the Fig. 6.5.

In the heat exchanger studied, ultrasound is applied to the shell (external tube), keeping hot water to flow in central pipe while cold water is made to flow in the annular space. The study has been carried out for parallel and counter flow and effects of ultrasound on overall heat transfer coefficient have been reported as

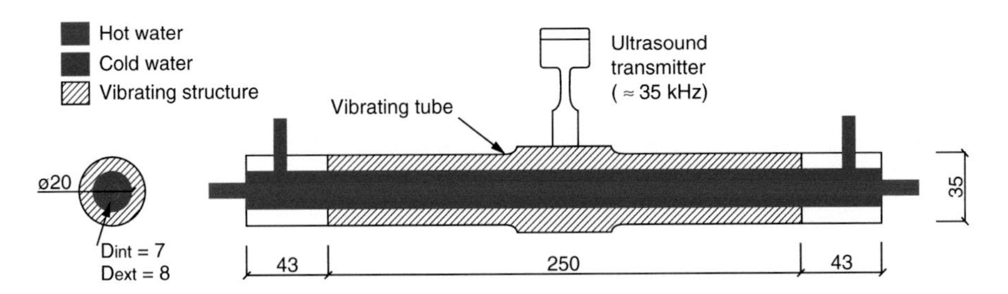

Fig. 6.5 Sketch and dimensions of the vibrating double pipe heat exchanger: dimensions in millimeters

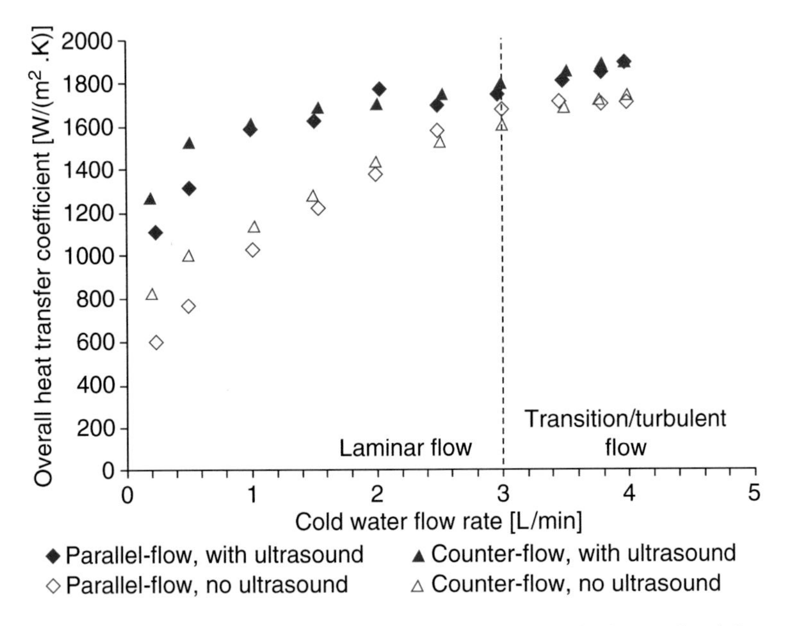

Fig. 6.6 Overall heat transfer coefficient versus cold water flow rate in the annulus: influence of flow configuration (hot water flow rate fixed: 1.6 L/min)

shown in Fig. 6.6. From this, following useful information may be noted: (1) ultrasound effect is much more intense on the shell side fluid than centre pipe fluid (2) as long as the cold water flow is laminar, there is an improvement in overall heat transfer coefficient due to application of high power ultrasound but as the flow becomes transient/turbulent, the improvement reduces (3) for same flow rate of cold water, there is no significant difference in the value of overall heat transfer coefficient for parallel and counter flow configurations.

It is important to note that since the experiment was conducted under forced convection situation, so out of two dominating causes (acoustic streaming and acoustic cavitations), there is no contribution of acoustic streaming in heat transfer enhancement; rather, cavitation bubbles implosion as well as vibrations of the walls result in a disturbance of the dynamic boundary layer and is the major cause of heat transfer augmentation.

Chapter 7
Conclusions

Abstract The advances in heat transfer enhancement techniques in all its aspects have been dealt with.

Keywords Heat transfer enhancement techniques • Active technique • Passive technique • Combined techniques • Heat transfer coefficient and friction factor correlations • Performance evaluation criteria

Energy crisis and environmental issues are the prime factors giving compulsory drive to improve the performance of heat exchangers that are source of major power consumption and ultimately affecting plant operating cost. In order to address the performance issues of heat exchangers, researchers have proposed different active, passive and compound techniques. Among these, active techniques have gained wide popularity in the domain of medical instruments, spacecraft engineering, marine applications and last but not the least all domains where performance is important but not the cost. Needless to say, apart from the above mentioned domain, a low cost solution to improve heat exchanger problem is highly desirable. Hence active approach of improving performance is not preferred in industries due to its high implementation and operating cost. Passive techniques where flow is modified either by altering the channel geometry or using insert devices, offers low cost solution to improve heat exchange process. Compound techniques, as discussed in this book, is the most recent techniques of improving heat exchange process and through this approach it would be possible to further reduce the size of the heat exchangers without compromising much the thermohydraulic performance. With the advent of nano technology, it is further possible to improve the performance of heat exchangers, particularly dealing with low thermal conductivity fluids. Although it has been reported by researchers that addition of nano particles significantly increases the pressure drop, yet this new era carries many possibility to contribute.

The advances in heat transfer enhancement techniques in all its aspects have been dealt with. The part I has dealt with only single-phase flow. The two-phase flow will be presented in the part II of the Book.

S.K. Saha et al., *Advances in Heat Transfer Enhancement*,
DOI 10.1007/978-3-319-29480-3_7

Bibliography

Abdullah MK, Abdullah MZ, Ramana MV, Khor Mujeebu CY, Ooi Y, Ripin ZM (2009) Numerical and experimental investigations on effect of fan height on the performance of piezoelectric fan in microelectronic cooling. Int Commun Heat Mass Transfer 36(1):51–58

Acikalin T, Garimella SV (2009) Analysis and prediction of the thermal performance of piezoelectrically actuated fans. Heat Transfer Eng 30(6):487–498

Acikalin T, Wait S, Garimella S, Raman A (2004) Experimental investigation of the thermal performance of piezoelectric fans. Heat Transfer Eng 25(1):4–14

Acikalin T, Garimella SV, Raman A, Petrosk J (2007) Characterization and optimization of the thermal performance of miniature piezoelectric fans. Int J Heat Fluid Flow 28(4, SI):806–820 (International conference on modelling fluid flow, Budapest, 2006)

Agostini B, Fabbri M, Park JE (2007) State of the art of high heat flux cooling technologies. Heat Transfer Eng 28(4):258–281

Allen P, Cooper P (1987) The potential of electrically enhanced evaporators. In: Third international symposium on the large scale application of heat pumps. pp 221–229

Allen P, Karayiannis T (1995) Electrohydrodynamic enhancement of heat transfer and fluid flow. Heat Recovery Syst CHP 15(5):389–423

Alzuaga S, Ballandras S, Bastien F, Daniau W, Gauthier-Manuel B, Manceau JF, Cretin B, Vairac P, Laude V, Khelif A, Duhamel R (2003) A large scale X-Y positioning and localisation system of liquid droplet using SAW on LiNbO3, IEEE symposium on ultrasonics, vol 2. IEEE, pp 1790–1793

Alzuaga S, Manceau J, Bastien F (2005) Motion of droplets on solid surface using acoustic radiation pressure. J Sound Vib 282(1–2):151–162

Arik M, Petroski J, Bar-Cohen A, Demiroglu M (2008) Energy efficiency of low form factor cooling devices. In: Proceedings of the ASME international mechanical engineering congress and exposition 2007 – heat transfer, fluid flows and thermal systems, pp 1347–1354

ASME (1998) ASME boiler and pressure vessel code, Sec. VIII, Div. 1, rules for construction of pressure vessels. American Society of Mechanical Engineers, New York

Bahrami M, Yovanovich MM, Culham JR (2006) Pressure drop of fully-developed, laminar flow in microchannel of arbitrary cross-section. J Fluids Eng 128(3):632–637

Bar-Cohen A, Jelinek M (1985) Optimum arrays of longitudinal, rectangular fins in convective heat transfer. Heat Transfer Eng 6(3):68–78

Bayraktar T, Pidugu SB (2006) Characterization of liquid flows in microfluidic systems. Int J Heat Mass Transf 49(5):815–824

Bejan A, Morega AM (1993) Optical arrays of pin fins and plate fins in laminar forced convection. ASME J Heat Transf 115(1):75–81

Benjamin KC (2002) Recent avances in 1–3 piezoelectric composite transducer technology for AUV/UUV acoustic imaging applications. J Electroceram 8(2):145–154

Bergles A (1964) In influence of flow vibration on forced-convection. J Heat Transf 107:559–560

Bergles E (1969) Survey and evaluation of techniques to augment convective heat and mass transfer, vol 1. Pergamon, Oxford

Bergles AE (1978) Enhancement of heat transfer. In: Heat transfer 1978, proceedings of the 6th international heat transfer conference, vol 6. Hemisphere, Washington, DC, pp 89–108

Bergles AE (1988) The role of experimentation in thermo-fluid sciences. In: Experimental heat transfer, fluid mechanics and thermodynamics 1988. Elsevier, New York, pp 676–684

Bergles AE (1997) Heat transfer enhancement—the encouragement and accommodation of high heat fluxes. J Heat Transf 119:8–19

Bergles AE (1998) Techniques to enhance heat transfer. In: Handbook of heat transfer, 3rd edn. McGraw-Hill, New York, pp 11.11–11.76

Bergles A (1999) Enhanced heat transfer: endless frontier, or mature and routine? J Enhanc Heat Transfer 6(2–4):79–88

Bergles AE, Kandlikar SG (2005) On the nature of critical heat flux in microchannels. ASME J Heat Transf 127(1):101–107

Bergles AE, Newell PH (1965) The influence of ultrasonic vibrations on heat transfer to water flowing in annuli. Int J Heat Mass Transf 8(10):1273–1280

Bergles AE, Blumenkrantz AR, Taborek J (1974) Performance evaluation criteria for enhanced heat transfer surfaces. In: Heat transfer 1974, proceedings of the 5th international heat transfer conference, vol II. The Japan Society of Mechanical Engineers, Tokyo, pp 234–238

Bergles AE, Jensen MK, Shome B (1996) Bibliography on enhancement of convective heat and mass transfer, RPI heat transfer laboratory report HTL-23, 1995. Preface in J Enhanc Heat Transfer 4:1–6

Bergman TL (2009) Effect of reduced specific heats of nanofluids on single phase, laminar internal forced convection. Int J Heat Mass Transf 52(5–6):1240–1244

Bhattacharyya S, Pal A, Nath G (1996) Unsteady flow and heat transfer between rotating coaxial disks. Numer Heat Transfer Part A Appl 30(5):519–532. doi:10.1080/10407789608913854

Bhogare RA, Kothawale BS (2013) A review on applications and challenges of nano-fluids as coolant in automobile radiator. Int J Sci Res Publica 3(8)

Bilen K, Cetin M, Gul H, Balta T (2009) The investigation of groove geometry effect on heat transfer for internally grooved tubes. Appl Therm Eng 29:753–761

Bochirol L, Bonjour E, Weil L (1960) Etude de l'action de champs électriques sur les transferts de chaleur dans les liquides bouillants. C R Hebd Seances Acad Sci 250(1):76–78

Bologa M, Savin I, Didkovsky A (1987) Electric field-induced enhancement of vapor condensation heat transfer in the presence of non-condensable gas. Int J Heat Mass Transf 30(8):1577–1585

Bologa M, Sajin T, Kozhukhar L, Klimov S, Motorin O (1996) The influence of electric fields on basic processes connected with physical phenomena in twophase systems. In: International conference on conduction and breakdown in dielectric liquid, pp 69–72

Bologa M, Savin I, Didkovsky A (1997) Enhancement of heat transfer in film condensation of vapours of dielectric liquids by superposition of electric fields. Heat Transfer Soviet Res 9(1):147–151

Bontemps A, Garrigue A, Goubier C, Huetz J, Marvillet C, Mercier P, Vidil R (1994) Echangeur de chaleur: intensification des échanges thermiques. Techniques de l'ingénieur

Brackbill TP, Kandlikar SG (2010) Application of lubrication theory and study of roughness pitch during laminar, transition, and low Reynolds number turbulent flow at microscale. Heat Transfer Eng 31(8):635–645

Bryszewska-Mazurek A, Mazurek W (2009) The influence of electric field on HFC- 245fa condensation. Mater Sci Pol 27(4):1257–1261

Butrymowicz D, Trela M, Karwacki J (2002) Enhancement of condensation heat transfer by means of EHD condensate drainage. Int J Therm Sci 41(7):646–657 (5th world conference on experimental heat transfer, fluid mechanics and thermodynamics, Thessaloniki, 24–28 Sept 2001)

Cai J, Huai X, Yan R, Cheng Y (2009) Numerical simulation on enhancement of natural convection heat transfer by acoustic cavitation in a square enclosure. Appl Therm Eng 29(10):1973–1982

Cardella A, Celata GP, Dell'Orco G, Gaspari G, Cattadori G, Mariani A (1992) Thermal hydraulics experiments for the net divertor. In: Proceedings of the 17th symposium on fusion technology, vol 1. pp 206–210

Cavallini A, Del Col D, Doretti L, Longo G, Rossetto L (1999) A new computational procedure for heat transfer and pressure drop during refrigerant condensation inside enhanced tubes. J Enhanc Heat Transfer 6(6):441–456

Celata GL (2004) Series in thermal and fluid physics and engineering. Begell House, New York

Celata GP, Cumo M, Mcphail S, Zummo G (2006) Characterization of fluid dynamic behavior and channel wall effects in microtube. Int J Heat Fluid Flow 27(1):135–143

Champagne PR, Bergles AE (2001) Development and testing of a novel, variable roughness technique to enhance, on demand, heat transfer in a single-phase heat exchanger. J Enhanc Heat Transfer 8:341–352

Chandrasekar M, Suresh S (2009) A review on the mechanisms of heat transport in nanofluids. Heat Transfer Eng 30(14):1136–1150

Chandratilleke T, Jagannatha D, Narayanaswamy R (2010) Heat transfer enhancement in micro-channels with cross-flow synthetic jets. Int J Therm Sci 49(3):504–513

Chang SW, Liou T-M, Juan W-C (2005) Influence of channel height on heat transfer augmentation in rectangular channels with two opposite rib-roughened walls. Int J Heat Mass Transf 48(13):2806–2813

Chang S, Yang T, Liou JS (2007) Heat transfer and pressure drop in tube with broken twisted tape insert. Exp Thermal Fluid Sci 32:489–501

Chaudhari M, Puranik B, Agrawal A (2010a) Heat transfer characteristics of synthetic jet impingement cooling. Int J Heat Mass Transf 53(5–6):1057–1069

Chaudhari M, Puranik B, Agrawal A (2010b) Effect of orifice shape in synthetic jet based impingement cooling. Exp Thermal Fluid Sci 34(2):246–256

Chaudhari M, Puranik B, Agrawal A (2011) Multiple orifice synthetic jet for improvement in impingement heat transfer. Int J Heat Mass Transf 54(9–10):2056–2065

Chein R, Chuang J (2007) Experimental microchannel heat sink performance studies using nanofluids. Int J Therm Sci 46(1):57–66

Chein RY, Huang GM (2005) Analysis of microchannel heat sink performance using nanofluids. Appl Therm Eng 25(17–18):3104–3114

Chen R, Chow L, Navedo J (2002) Effects of spray characteristics on critical heat flux in subcooled water spray cooling. Int J Heat Mass Transf 45(19):4033–4043

Chen R, Chow L, Navedo J (2004) Optimal spray characteristics in water spray cooling. Int J Heat Mass Transf 47(23):5095–5099

Chen F, Peng Y, Song Y, Chen M (2007) EHD behavior of nitrogen bubbles in DC electric fields. Exp Thermal Fluid Sci 32(1):174–181

Cheng L, Bandarra EP, Thome JR (2008) Nanofluid two-phase flow and thermal physics: a new research frontier of nanotechnology and its challenges. J Nanosci Nanotechnol 8(7):3315–3332

Cheng W-L, Liu Q-N, Zhao R, Fan H-l (2010) Experimental investigation of parameters effect on heat transfer of spray cooling. Heat Mass Transf 46(8–9):911–921

Cheung K, Ohadi M, Dessiatoun S (1999) EHD-assisted external condensation of R- 134a on smooth horizontal and vertical tubes. Int J Heat Mass Transf 42(10):1747–1755

Choi H (1968) Electrohydrodynamic condensation heat transfer. J Heat Transf 90(1):98–102

Choi SUS (2009) Nanofluids: from vision to reality through research. J Heat Transf 131:033106-1

Choi SUS, Jang SP (2006) Cooling performance of a microchannel heat sink with nanofluids. Appl Therm Eng 26(17–18):2457–2463

Choi H, Reynolds M (1965) Study of electrostatic effects on condensing heat transfer. Tech. rep., Air Force technical report

Colgan EG, Furman B, Gaynes M, Graham W, Labianca N, Magerlein JH, Polastre RJ, Rothwell MB, Bezama RJ, Choudhary R, Marston K, Toy H, Wakil J, Zitz J, Schmidt R (2005) A practical implementation of silicon microchannel coolers for high power chips. In: Proceedings of the 21st annual IEEE semiconductor thermal measurement and management symposium, San Jose, 15–17 March 2005. pp 1–7

Consolini L, Thome JR (2010) A heat transfer model for evaporation of coalescing bubbles in micro-channel flow. Int J Heat Fluid Flow 31(1):115–125

Cooper P, Allen P (1984) The potential of electrically enhanced condensers. In: Proceedings second international symposium in the large scale application of heat pumps. pp 295–309

Cotton JS (2009) Electrohydrodynamic condensation heat transfer modulation under DC and AC applied voltages in a horizontal annular channel. IEEE Trans Dielectr Electr Insul 16(2):495–503

Croce G, D'agaro P, Nonino C (2007) Three-dimensional roughness effect on microchannel heat transfer and pressure drop. Int J Heat Mass Transf 50(25–26):5249–5259

Da Silva L, Molki M, Ohadi M (2000) Electrohydrodynamic enhancement of R-134a condensation on enhanced tubes. In: IAS 2000 – conference record of the 2000 IEEE industry applications conference, pp 757–764
Daaboul M (2009) Etude et développement d'actionneurs électrohydrodynamiques pour le contrôle des écoulements. Application à l'atomisation des nappes liquides, Ph.D. thesis, Université de Poitiers Sciences et Ingénierie en Matériaux, Mécanique, Energétique et Aéronautique
Dalkilic AS, Wongwises S (2009) Intensive literature review of condensation inside smooth and enhanced tubes. Int J Heat Mass Transf 52(15–16):3409–3426
Damianidis C, Collins M, Karayiannis T, Allen P (1990) EHD effect in condensation of dielectric fluid. In: Proceedings second international symposium ion condensers and condensation, Bath. pp 505–518
Das SK, Choi SUS, Patel HE (2006) Heat transfer in nanofluids— a review. Heat Transfer Eng 27(10):3–19
Daungthongsuk W, Wongwises S (2007) A critical review of convective heat transfer of nanofluids. Renew Sustain Energy Rev 11(5):797–817
Debbaut B (2001) Non-isothermal and viscoelastic effects in the squeeze flow between infinite plates. J Non-Newtonian Fluid Mech 98(1):15–31. doi:10.1016/S0377-0257(01)00096-9
Decarpigny J, Hamonic B, Wilson OJ (1991) The design of low-frequency underwater acoustic projectors: present status and future trends. IEEE J Ocean Eng 16(1):107–122
Dewan A, Mahanta P, Raju KS, Kumar PS (2004) Review of passive heat transfer augmentation techniques. J Power Energy 218:509–525
Dey S, Chakrborty D (2009) Enhancement of convective cooling using oscillating fins. Int Commun Heat Mass Transfer 36(5):508–512
Didkovsky A, Bologa M (1981) Vapor film condensation heat transfer and hydrodynamics under the influence of an electric field. Int J Heat Mass Transf 24(5):811–819
Eames I, Sabir H (1997) Potential benefits of electrohydrodynamic enhancement of two-phase heat transfer in the design of refrigeration systems. Appl Therm Eng 17(1):79–92
Earls Brennen C (1995) Cavitation and bubbles dynamics. Oxford University Press, New York
Egan E, Amon CH (2000) Thermal management strategies for embedded electronic components of wearable computers. J Electron Packag 122:98
Eiamsa-ard S, Promvonge P (2010) Thermal characteristics in round tube fitted with serrated twisted tape. Appl Therm Eng 30:1673–1682
Eiamsa-ard S, Seemawute P, Wongcharee K (2010a) Influences of peripherally-cut twisted tape insert on heat transfer and thermal performance characteristics in laminar and turbulent tube flows. Exp Thermal Fluid Sci 34:711–719
Eiamsa-ard S, Nivesrangsan P, Chokphoemphun S, Promvonge P (2010b) Influence of combined non-uniform wire coil and twisted tape inserts on thermal performance characteristics. Int Commun Heat Mass Transfer 37:850–856
Estes K, Mudawar I (1995) Correlation of Sauter mean diameter and critical heat flux for spray cooling of small surfaces. Int J Heat Mass Transf 38(16):2985–2996
Fan Y, Luo LG (2008) Recent applications of advances in microchannel heat exchangers and multi-scale design optimization. Heat Transfer Eng 29(5):461–474
Fan JF, Ding WK, Zhang JF, He YL, Tao WQ (2009) A performance evaluation plot of enhanced heat transfer techniques oriented for energy-saving. Int J Heat Mass Transf 52:33–44
Fand R (1965) Influence of acoustic vibrations on heat transfer by natural convection from a horizontal cylinder to water. J Heat Transf 87(2):309–310
Fedorov AG, Viskanta R (2000) Three-dimensional conjugate heat transfer in the microchannel heat sink for electronic packaging. Int J Heat Mass Transf 43(3):399–415
Feng Y, Seyed-Yagoobi J (2002) Linear instability analysis of a horizontal two-phase flow in the presence of electrohydrodynamic extraction force. J Heat Transfer Trans ASME 124(1): 102–110
Fernandez J, Poulter R (1987) Radial mass-flow in electrohydrodynamicallyenhanced forced heat transfer in tubes. Int J Heat Mass Transf 30(10):2125–2136

Fiebig M, Mitra NK (eds) (1998) Vortices and heat transfer. Vieweg, Braunschweig
Florio LA, Harnoy A (2007a) Combination technique for improving natural convection cooling in electronics. Int J Therm Sci 46(1):76–92
Florio LA, Harnoy A (2007b) Augmenting natural convection in a vertical flow path through transverse vibrations of an adiabatic wall. Numer Heat Trans Part A Appl 52(6):497–530
Florio LA, Harnoy A (2011) Natural convection enhancement by a discrete vibrating plate and a cross-flow opening: a numerical investigation. Heat Mass Transf 47(6):655–677
Foong AJL, Ramesh N, Chandratilleke TT (2009) Laminar convective heat transfer in a microchannel with internal longitudinal fins. Int J Therm Sci 48(10):1908–1913
Gachagan A, Speirs D, McNab A (2003) The design of a high power ultrasonic test cell using finite element modelling techniques. Ultrasonics 41(4):283–288
Gachagan A, McNab A, Blindt R, Patrick M, Marriott C (2004) A high power ultrasonic array based test cell. Ultrasonics 42(1–9):57–68
Gamrat G, Favre-Marinet M, Le Person S (2009) Modelling of roughness effects on heat transfer in thermally fully-developed laminar flows through microchannels. Int J Therm Sci 48(12):2203–2214
Garcia A, Solano J, Vicente P, Viedma A (2007) Enhancement of laminar and transitional flow heat transfer in tubes by means of wire coil inserts. Int J Heat Mass Transf 15–16(50):3176–3189
Gardon R, Carbonpue J (1962) Heat transfer between a flat plate and jets of air impinging on it. In: International developments in heat transfer, second international heat transfer conference. ASME, pp 454–460
Garimella SV, Singhal V (2004) Single-phase flow and heat transport and pumping considerations in microchannel heat sinks. Heat Transfer Eng 25(1):15–25
Garrity PT, Klausner JF, Mei R (2009) Instability phenomena in a two-phase microchannel thermosyphon. Int J Heat Mass Transf 52(7–8):1701–1708
Gauntner JWW, Livingood JNB, Hrycak P (1970) Survey of literature on flow characteristics of a single turbulent jet impinging on a flat plate. Tech. rep., NASA TN D-5652 NTIS N70-18963
Gidwani A, Molki M, Ohadi M (2002) EHD-enhanced condensation of alternative refrigerants in smooth and corrugated tubes. HVAC&R Res 8(3):219–237
Goodling JS (1993) Microchannel heat exchangers: a review. In: Proceedings of the high heat flux engineering II, San Diego, 1997, 12–13 July 1993. pp 66–82
Guo Z-Y, Li Z-X (2003) Size effect on single-phase channel flow and heat transfer at microscale. Int J Heat Fluid Flow 24(3):284–298
Hamza E (1992) Unsteady flow between two disks with heat transfer in the presence of a magnetic field. J Phys D Appl Phys 25(10):1425–1431. doi:10.1088/0022-3727/25/10/007
Harris C, Despa M, Kelly K (2000) Design and fabrication of a cross flow micro heat exchanger. J Microelectromech Syst 9(4):502–508
Hashimoto Y (1998) Transporting objects without contact using flexural traveling waves. J Acoust Soc Am 103:3230
Hassan I, Phutthavong P, Abdelgawad M (2004) Microchannel heat sinks: an overview of the state-of-the-art. Microscale Thermophys Eng 8(3):183–205
Havelock D, Kuwano S, Vorlander M (eds) (2008) Handbook of signal processing in acoustics. Springer, New York
Heris S, Etemad S, Esfahany A (2006) Experimental investigation of oxidenanofluids laminar flow convective heat transfer. Int Commun Heat Mass Transfer 33(4):529–535
Hetsroni G, Mosyak A, Segal Z (2000) Nonuniform temperature distribution in electronic devices cooled by flow in parallel microchannels. IEEE Trans Compon Packag Technol 24(1):16–23
Hetsroni G, Mosyak A, Segal Z, Pogrebnyak E (2003) Two-phase flow patterns in parallel microchannels. Int J Multiphase Flow 29(3):341–360
Hewitt GF, Shires GL, Bott TR (1994) Process heat transfer. CRC, Boca Raton
Holmes R, Chapman A (1970) Condensation of Freon-114 in presence of a strong nonuniform, alternating electric field. J Heat Transf 92(4):616–621
Horacek B, Kiger K, Kim J (2005) Single nozzle spray cooling heat transfer mechanisms. Int J Heat Mass Transf 48(8):1425–1438

Hu YD, Werner C, Li DQ (2003) Influence of three-dimensional roughness on pressure-driven flow through microchannels. J Fluids Eng 125(5):871–879

Hyun S, Lee D, Loh B (2005) Investigation of convective heat transfer augmentation using acoustic streaming generated by ultrasonic vibrations. Int J Heat Mass Transf 48(3–4):703–718

Ihara A, Watanabe H (1994) On the flow around flexible plates, oscillating with large-amplitude. J Fluids Struct 8(6):601–619

Jachuck R (1999) Opportunities presented by cross corrugated polymer film compact heat exchangers. In: Compact heat exchangers and enhancement technology for the process industries. Begell House, New York, pp 243–250

Jacobi AM, Shah RK (1998) Air-side flow and heat transfer in compact heat exchangers: a discussion of enhancement mechanisms. Heat Transfer Eng 19(4):29–41

Jacobi AM, Thome JR (2002) Heat transfer model for evaporation of elongated bubble flows in microchannels. ASME J Heat Transf 124(6):1131–1136

Jalaluddin A, Sinha D (1962) Effect of an electric field on the superheat of liquids. Nuovo Cimento 26Ser X:234–237

Jambunathan K, Lai E, Moss M, Button B (1992) A review of heat transfer data for single circular jet impingement. Int J Heat Fluid Flow 13(2):106–115

Jensen M (1985) An evaluation of the effect of twisted tape swirl generator in twophase flow heat exchangers. Heat Transfer Eng 6:19

Jiang L, Wong M, Zohar Y (1999) Phase change in microchannel heat sinks with integrated temperature sensors. J Microelectromech Syst 8(4):358–365

Jia-Xiang Y, Li-Jian D, Yang H (1996) An experimental study of EHD coupled heat transfer. IEEE 1:348–351

Judy J, Maynes D, Webb BW (2002) Characterization of frictional pressure drop for liquid flows through microchannels. Int J Heat Mass Transf 45(17):3477–3489

Jung J-Y, Oh H-S, Kwak H-Y (2009) Forced convective heat transfer of nanofluids in microchannels. Int J Heat Mass Transf 52(1–2):466–472

Kakac S, Pramuanjaroenkij A (2009) Review of convective heat transfer enhancement with nanofluids. Int J Heat Mass Transf 52(13–14):3187–3196

Kakaç S, Bergles AE, Mayinger F, Yüncü H (eds) (1999) Heat transfer enhancement of heat exchangers. Kluwer, Dordrecht

Kalinin E, Dreitser G (1998) Heat transfer enhancement in heat exchangers advances. Heat Transfer 31:159–332

Kamkari B, Alemrajabi AA (2010) Investigation of electrohydrodynamicallyenhanced convective heat and mass transfer from water surface. Heat Transfer Eng 31(2):138–146

Kandlikar SG (2003) Methods for stabilizing flow in channels and systems thereof. US patent application no. 20090266436

Kandlikar SG (2005) Roughness effects at microscale—reassessing Nikuradse's experiments on liquid flow in rough tubes. Bull Pol Acad Sci Tech Sci 53(4):343–349

Kandlikar SG (2008) Exploring roughness effect on laminar internal flow—are we ready for change? Nanoscale Microscale Thermophys Eng 12(1):61–82

Kandlikar SG (2010) Microchannels: rapid growth of a nascent technology. ASME J Heat Transf 132(4):040301

Kandlikar SG, Bapat AV (2007) Evaluation of jet impingement, spray and microchannel chip cooling options for high heat flux removal. Heat Transfer Eng 28(11):911–923

Kandlikar SG, Grande WJ (2003) Evolution of microchannel flow passages—thermohydraulic performance and fabrication technology. Heat Transfer Eng 24(1):3–17

Kandlikar SG, Grande WJ (2004) Evaluation of single phase flow in microchannels for high heat flux chip cooling-thermohydraulic performance enhancement and fabrication technology. Int J Therm Sci 45(11):1073–1083

Kandlikar SG, Upadhye HR (2005) Extending the heat flux limit with enhanced microchannels in dorect single-phase cooling of computer chips. In: Proceedings of twenty-first annual IEEE semiconductor thermal measurement and management symposium, 2005 March 15–17. pp 8–15

Kandlikar SG, Joshi S, Tian S (2003) Effect of surface roughness on heat transfer and fluid flow characteristics at low Reynolds numbers in small diameter tubes. Heat Transfer Eng 24(3):4–16

Kandlikar SG, Schmitt D, Carrano AL, Taylor JB (2005a) Characterization of surface roughness effects on pressure drop in single- phase flow in minichannels. Phys Fluids 17(10):100606

Kandlikar SG, Kuan WK, Mukherjee A (2005b) Experimental study of heat transfer in an evaporating meniscus on a moving heated surface. ASME J Heat Transf 127(3):244–252

Kandlikar SG, Garimella S, Li D, Colin S, King MR (2006) Heat transfer and fluid flow in minichannels and microchannels. Elsevier, Kidlington

Kanizawa F, Ribatski G (2011) Two-phase flow patterns and frictional pressure drop inside tubes containing twisted-tape insert. In: The 6th international conference on transport phenomena in multiphase systems

Karayiannis T, Allen P (1991) Electro-hydrodynamic enhancement of two-phase heat transfer. In: Eurotech Direct Congress thermofluids engineering. pp 165–181

Keblinski P, Phillpot SR, Choi SUS, Eastman JA (2002) Mechanisms of heat flow in suspensions of nano-sized particles (nanofluids). Int J Heat Mass Transf 45(4):855–863

Khaled A, Vafai K (2003) Analysis of flow and heat transfer inside oscillatory squeezed thin films subject to a varying clearance. Int J Heat Mass Transf 46(4):631–641. doi:10.1016/S0017-9310(02)00328-9

Kim J (2007) Spray cooling heat transfer: the state of the art. Int J Heat Fluid Flow 28 (4, Sp Iss SI):753–767

Kim M-H, Lee S, Mehendale S, Webb RL (2003) Advances in heat transfer. Elsevier, San Diego

Kimber M, Garimella SV (2009a) Cooling performance of arrays of vibrating cantilevers. J Heat Transfer Trans ASME 131(11):1–8

Kimber M, Garimella SV (2009b) Measurement and prediction of the cooling characteristics of a generalized vibrating piezoelectric fan. Int J Heat Mass Transf 52(19–20):4470–4478

Kimber M, Garimella SV, Raman A (2007) Local heat transfer coefficients induced by piezoelectrically actuated vibrating cantilevers. J Heat Trans Trans ASME 129(9):1168–1176

Kimber M, Suzuki K, Kitsunai N, Seki K, Garimella SV (2008) Quantification of piezoelectric fan flow rate performance and experimental identification of installation effects. In: 11th IEEE intersociety conference on thermal and thermomechanical phenomena in electronic systems, pp 471–479

Kirby BJ (2010) Micro- and nanoscale fluid mechanics. Cambridge University, New York

Kishimoto T, Sasaki S (1987) Cooling characteristics of diamond-shaped interrupted cooling fin for high-power LSI devices. Electron Lett 23(9):456–457

Kleinstreuer C, Koo J (2004) Computational analysis of wall roughness effects for liquid flow in micro-conduits. J Fluids Eng 126(1):1–9

Knight RW, Hall DJ, Goodling JS, Jaeger RC (1992) Heat sink optimization with application to microchannels. IEEE Trans Compon Hybrids Manuf Technol 15(5):832–842

Koo J, Kleinstreuer C (2005) Laminar nanofluid flow in microheat-sinks. Int J Heat Mass Transf 48(13):2652–2661

Kosar A, Peles Y (2006a) Thermal-hydraulic performance of MEMS-based pin fin heat sink. ASME J Heat Transf 128(2):121–131

Kosar A, Peles Y (2006b) Convective flow of refrigerant (R-123) across a bank of micro pin fins. Int J Heat Mass Transf 49(17–18):3142–3155

Kosar A, Peles Y (2007) Critical heat flux of R-123 in silicon-based microchannels. ASME J Heat Transf 129(7):844–851

Kosar A, Peles Y, Bergles AE, Cole GS (2009) Experimental investigation of critical heat flux in microchannels for flow-field probes. Paper no. ICNMM2009-82214, ASME seventh international conference on nanochannels, microchannels, and minichannels, Pohang, 22–24 June

Kumar P, Topin F, Miscevic M, Lavieille P, Tadrist L (2012) Heat transfer enhancement in short corrugated mini-tubes, numerical heat and mass transfer in porous media. Adv Struct Mater 27:181–208

Langlois W (1962) Isothermal squeeze films. Q Appl Math 131–150
Laohalertdecha S, Wongwises S (2006) Effects of EHD on heat transfer enhancement and pressure drop during two-phase condensation of pure R-134a at high mass flux in a horizontal micro-fin tube. Exp Thermal Fluid Sci 30:675–686
Laohalertdecha S, Wongwises S (2007a) Effect of EHD on heat transfer enhancement during two-phase condensation of R-134a at high mass flux in a horizontal smooth tube. Heat Mass Transf 43(9):871–880
Laohalertdecha S, Wongwises S (2007b) A comparison of the effect of the electrohydrodynamic technique on the condensation heat transfer of HFC- 134a inside smooth and micro-fin tubes. J Mech Sci Technol 21(12):2168–2177
Laohalertdecha S, Naphon P, Wongwises S (2007) A review of electrohydrodynamic enhancement of heat transfer. Renew Sustain Energy Rev 11(5):858–876
Leal L (2012) Etude des mécannismes de nucléation par action simultanée de l'ébullition et de la cavitation. Ph.D. thesis, Université de Toulouse, Ecole doctorale Mécanique, Energétique, Génie Civil, Procédés
Lee PS, Garimella SV (2006) Thermally developing flow and heat transfer in rectangular microchannels of different aspect ratios. Int J Heat Mass Transf 49(17–18):3060–3067
Lee J, Mudawar I (2007) Assessment of the effectiveness of nanofluids for single-phase and two-phase heat transfer in micro-channels. Int J Heat Mass Transf 50(3–4):452–463
Lee S, Choi SUS, Li S, Eastman JA (1999) Measuring thermal conductivity of fluids containing oxide nanoparticles. ASME J Heat Transf 121(2):280–289
Lee P-S, Garimella V, Dong L (2005) Investigation of heat transfer in rectangular microchannels. Int J Heat Mass Transf 48(9):1688–1704
Legay M, Gondrexon N, Le Person S, Boldo P, Bontemps A (2011) Enhancement of heat transfer by ultrasound: review and recent advances. Int J Chem Eng 2011:1–17, Hindawi Publishing Corporation
Legay M, Simony B, Boldo P, Gondrexon N, Le Person S, Bontemps A (2012) Improvement of heat transfer by means of ultrasound: application to a double-tube heat exchanger. Ultrason Sonochem 19:1194–1200
Lelea D, Nishio S, Takano K (2004) The experimental research on microtube heat transfer and fluid flow of distilled water. Int J Heat Mass Transf 47(12–13):2817–2830
Lemlich R, Hwu C (1961) The effect of acoustic vibration on forced convective heat transfer. AIChE J 7(1):102–106
Li K, Parker J (1967) Acoustical effects on free convective heat transfer from a horizontal wire. J Heat Transf 89(3):277–278
Li BQ, Cader T, Schwarzkopf J, Okamoto K, Ramaprian B (2006) Spray angle effect during spray cooling of microelectronics: experimental measurements and comparison with inverse calculations. Appl Therm Eng 26(16):1788–1795
Lienhard JH (1998) Review of heat transfer augmentation in turbulent flows. Appl Mech Rev 51(2):B19
Liu D, Garimella SV (2005) Analysis and optimization of the thermal performance of microchannel heat sinks. Int J Numer Methods Heat Fluid Flow 15(1):7–26
Liu S-F, Huang R-T, Sheu W-J, Wang C-C (2009) Heat transfer by a piezoelectric fan on a flat surface subject to the influence of horizontal/vertical arrangement. Int J Heat Mass Transf 52(11–12):2565–2570
Loh B, Hyun S, Ro P, Kleinstreuer C (2002) Acoustic streaming induced by ultrasonic flexural vibrations and associated enhancement of convective heat transfer. J Acoust Soc Am 111(2):875–883
Mala GM, Dongqing L (1999) Flow characteristics of water in microtubes. Int J Heat Fluid Flow 20(2):142–148
Mala GM, Dongqing L, Dale JD (1997) Heat transfer and fluid flow in microchannels. Int J Heat Mass Transf 40(13):3079–3088
Manglik RM, Bergles AE Heat transfer and pressure drop correlations for twisted-tape inserts in isothermal tubes part I laminar flows. J Heat Transf

Manglik R, Bergles AE (1993) Heat transfer and pressure drop correlations for twisted-tape inserts in isothermal tubes part II transition and turbulent flows. J Heat Transf 115:890
Manglik RM, Bergles AE (2003) Swirl flow heat transfer and pressure drop with twisted-tape inserts. Adv Heat Tran 36:183
Marshall WJ (1999) Underwater sound projectors. In: Webster JG (ed) Wiley encyclopedia of electrical and electronics engineering. Wiley, Hoboken
J.E. Martin, Underwater transducer and projector therefor, Patent US 3,992,693, 1976
Mason WP (1976) Sonics and ultrasonics: early history and applications. IEEE Trans Sonics Ultrason 23(4):224–231
Mays DC, Campbell AB, Nair SS, Mills JB, Thomas GA (2001) Thermal technology for space suits. ASHRAE J 43:25–36
Mehendale SS, Jacobi AM, Shah RK (2000) Fluid flow and heat transfer at micro- and meso-scales with application to heat exchanger design. Appl Mech Rev 53(7):175–193
Miner A, Ghoshal U (2006) Limits of heat removal in microelectronic systems. IEEE Trans Compon Packag Technol 29(4):743–749
Mishra C, Peles Y (2005a) Cavitation in flow through a micro-orifice inside a silicon microchannel. Phys Fluids 17(1):013601
Mishra C, Peles Y (2005b) Flow visualization of cavitating flows through a rectangular slot micro-orifice ingrained in a microchannel. Phys Fluids 17(11):113602–113614
Mishra C, Peles Y (2005c) Size scale effects on cavitating flows through microorifices entrenched in rectangular microchannels. J Microelectromech Syst 14(5):987–999
Mishra C, Peles Y (2006a) An experimental investigation of hydrodynamic cavitation in micro-Venturis. Phys Fluids 18(10):103603–103605
Mishra C, Peles Y (2006b) Development of cavitation in refrigerant (R-123) flow inside rudimentary microfluidic systems. J Microelectromech Syst 15(5):1319–1329
Missaggia LJ, Walpole JN, Liau ZL, Phillips RJ (1989) Microchannel heat sinks for two-dimensional high-power-density diode laser arrays. IEEE J Quantum Electron 25(9):1988–1992
Miyara A, Otsubo Y, Ohtsuka S, Mizuta Y (2003) Effects of fin shape on condensation in herringbone microfin tubes. Int J Refrig 26(4):417–424
Mori YK, Hijikata K, Hirasawa S, Nakayama W (1987) Optimized performance of condensers with outside condensing surfaces. J Heat Transf 103:96–102
Morini GL (2004) Single-phase convective heat transfer in microchannels: a review of experimental results. Int J Therm Sci 43(7):631–651
Morini GL, Lorenzini M, Salvigni S, Celata GP (2010) Experimental analysis of microconvective heat transfer in the laminar and transitional regions. Exp Heat Transfer 23(1):73–93
Moriyama K, Inoue A, Ohira H (1992) Thermohydraulic characteristics of two-phase flow in extremely narrow channels (the frictional pressure drop and void fraction of adiabatic two-component two-phase flow). Heat Transfer Jpn Res 21(8):823–837
Mudawar I, Estes K (1996) Optimizing and predicting CHF in spray cooling of a square surface. J Heat Transfer Trans ASME 118(3):672–679
Mundinger D, Beach R, Benett W, Solarz R, Krupke W, Staver R, Tuckerman D (1988) Demonstration of high-performance silicon microchannel heat exchangers for laser diode array cooling. Appl Phys Lett 53(12):1030–1032
Muralidhar Rao M, Sastri VMK (1995) Experimental investigation of fluid flow and heat transfer in a rotating tube with twisted tape inserts. Heat Transfer Eng 16(2):19–28
Murshed SMS, Leong KC, Yang C (2008) Thermophysical and electrokinetic properties of nanofluids—a critical review. Appl Therm Eng 28(17–18):2109–2125
Murugesan P, Mayilsamy K, Suresh S, Srinivasan PSS (2011) Heat transfer and pressure drop characteristics in a circular tube fitted with and without V-cut twisted tape insert. Int Commun Heat Mass Transfer 38:329–334
Nakamura M, Nakamura T, Tanaka T (2000) A computational study of viscous flow in a transversely oscillating channel. JSME Int J Ser C Mech Syst Mach Elem Manuf 43(4):837–844
Nanan K, Thianpong C, Promvonge P, Eiamsa-ar S (2014) Investigation of heat transfer enhancement by perforated helical twisted-tapes. Int Commun Heat Mass Transfer 52:106–112

Narasimha R, Sreenivasan KR (1979) Advances in applied mechanics. Elsevier, Amsterdam

Neve R, Yan Y (1996) Enhancement of heat exchanger performance using combined electrohydrodynamic and passive methods. Int J Heat Fluid Flow 17(4):403–409

Nicolay P (2007) Les capteurs à ondes élastiques de surface: applications pour la mesure des basses pressions et des hautes températures, thesis, Nancy 1

Nomura S, Yamamoto A, Murakami K (2002) Ultrasonic heat transfer enhancement using a horn-type transducer. Jpn J Appl Phys Part 1 – Regul Pap Short Notes Rev Pap 41(5B):3217–3222

Nusselt W (1916) Die Oberfl€achenkondensation des Wasserdampfes. Z Ver Dtsch Ing 60(27):541–546; 60(28):569–575

Obot NT (2002) Toward a better understanding of friction and heat/mass transfer in microchannels—a literature review. Microscale Thermophys Eng 6(3):155–173

Ohadi MM, Li SS, Dessiatoun S (1994) Electrostatic heat transfer enhancement in a tube bundle gas-to-gas heat exchanger. J Enhanc Heat Transfer 1:327–335

Ohadi MM, Dessiatoun SV, Darabi J, Salehi M (1996) Active augmentation of single-phase and phase-change heat transfer—an overview, in process. In: Enhanced and multiphase heat transfer. Begell House, New York, pp 271–286

Omidvarborna H, Mehrabani-Zeinabad A, Esfahany MN (2009) Effect of electrohydrodynamic (EHD) on condensation of R-134a in presence of noncondensable gas. Int Commun Heat Mass Transfer 36(3):286–291

Oyakawa K, Umeda A, Islam MD, Saji N, Matsuda S (2009) Flow structure and heat transfer of impingement jet. Heat Mass Transf 46(1):53–61

Ozerinc S, Kakac S, Yazicioglu AG (2010) Enhanced thermal conductivity of nanofluids: a state-of-the-art review. Microfluid Nanofluid 8(2):145–170

Pais M, Tilton D, Chow L, Mahefky E (1989) High heat flux, low superheat evaporative spray cooling. In: Proceedings of the 27th AIAA aerospace sciences meetings

Palm B (2001) Heat transfer in microchannels. Microscale Thermophys Eng 5(3):155–175

Palmer HJ (1976) The hydrodynamic stability of rapidly evaporating liquids at reduced pressure. J Fluid Mech 75(3):487–511

Park JE, Thome JR (2010) Critical heat flux in multi-microchannel copper elements with low pressure refrigerants. Int J Heat Mass Transf 53(1–3):110–122

Pate MB, Ayub ZH, Kohler J (1990) Heat exchangers for the air conditioning and refrigeration industry: state-of-the-art design and technology. In: Compact heat exchangers. Hemisphere, New York, pp 567–590

Peles YP, Kosar A, Mishra C, Kuo C-J, Schneider B (2005) Forced convective heat transfer across a pin fin micro heat sink. Int J Heat Mass Transf 48(17):3615–3627

Peng XF, Peterson GP, Wang BX (1994a) Frictional flow characteristics of water flowing through rectangular microchannels. Exp Heat Transfer 7(4):249–264

Peng XF, Peterson GP, Wang BX (1994b) Heat transfer characteristics of water flowing through microchannels. Exp Heat Transfer 7(4):265–283

Peyghambarzadeh SM, Hashemabadi SH, Seifi Jamnani M, Hoseini SM (2011) Improving the cooling performance of automobile radiator with Al2O3/water nanofluid. Appl Therm Eng 31:1833–1838

Phillips RJ (1987) Forced-convection, liquid-cooled, microchannel heat sinks. MSME Massachusetts Institute of Technology, Cambridge, MA

Phillips RJ (1990) Advances in thermal modeling of electronic components and systems. ASME, New York, Chap. 3

Phillips RJ, Glicksman L, Larson R (1987) Forced-convection, liquid-cooled, microchannel heat sinks for high-power-density microelectronics. In: Aung W (ed) Proceedings of the cooling technology for electronic equipment, Honolulu. pp 295–316

Posew K, Laohalertdecha S, Wongwises S (2009) Evaporation heat transfer enhancement of R-134a flowing inside smooth and micro-fin tubes using the electrohydrodynamic technique. Energy Convers Manag 50(7):1851–1861

Promvonge P (2008) Thermal augmentation in circular tube with twisted tape and wire coil turbulators. Energy Convers Manag 49(11):2949–2955

Promvonge P, Eiamsa-ard S (2006) Heat transfer enhancement in a tube with combined conical-nozzle inserts and swirl generator. Energy Convers Manag 47:2867–2882

Promvonge P, Eiamsa-ard S (2007a) Heat transfer behaviors in a tube with combined conical-ring and twisted-tape insert. Int Commun Heat Mass Transfer 34(7):849–859

Promvonge P, Eiamsa-ard S (2007b) Heat transfer augmentation in a circular tube using v-nozzle turbulator inserts and snail entry. Exp Thermal Fluid Sci 32(1):332–340

Promvonge P, Pethkool S, Pimsarn M, Thianpong C (2012) Heat transfer augmentation in a helical-ribbed tube with double twisted tape inserts. Int Commun Heat Mass Transfer 39:953–959

Qu WL, Mudawar I (2002) Experimental and numerical study of pressure drop and heat transfer in a single-phase micro-channel heat sink. Int J Heat Mass Transf 45(12):2549–2565

Qu W, Mudawar I (2004a) Transport phenomena in two-phase micro- channel heat sinks. J Electron Packag 126(2):213–224

Qu W, Mudawar I (2004b) Measurement and correlation of critical heat flux in two-phase micro-channel heat sinks. Int J Heat Mass Transf 47(10–11):2045–2059

Qu W, Mala GM, Li D (2000a) Pressure-driven water flows in trapezoidal silicon microchannels. Int J Heat Mass Transf 43(3):353–364

Qu W, Mala GM, Li D (2000b) Heat transfer for water flow in trapezoidal silicon microchannels. Int J Heat Mass Transf 43(21):3925–3936

Quéré D (2002) Surface chemistry – Fakir droplets. Nat Mater 1(1):14–15

Ravigururajan T, Bergles AE (1996) Development and verification of general correlations for pressure drop and heat transfer in single-phase turbulent flow in enhanced tubes. Exp Thermal Fluid Sci 13:55–70

Rawool AS, Mitra SK, Kandlikar SG (2006) Numerical simulation of flow through microchannels with designed roughness. Microfluid Nanofluid 2(3):215–221

Revellin R, Thome JR (2008) A theoretical model for the prediction of the critical heat flux in heated microchannels. Int J Heat Mass Transf 51(5–6):1216–1225

Roday AP, Jensen MK (2009) A review of the critical heat flux condition in mini- and microchannels. J Mech Sci Technol 23(9):2529–2547

Rolt KD (1990) History of the flextensional electroacoustic transducer. J Acoust Soc Am 87(3):1340–1349

Rosa P, Karayiannis TG, Collins MW (2009) Single-phase heat transfer in microchannels: the importance of scaling effects. Appl Therm Eng 29(17–18):3447–3468

Royne A, Dey CJ, Mills DR (2005) Cooling of photo voltaic cells under concentrated illumination: a critical review. Sol Energy Mater Sol Cells 86(4):451–483

Saha S (2010) Thermohydraulics of turbulent flow through rectangular and square ducts with axial corrugation roughness and twisted-tapes with and without oblique teeth. Exp Thermal Fluid Sci 34:744–752

Saha S, Saha SK (2013) Enhancement of heat transfer of laminar flow through a circular tube having integral helical rib roughness and fitted with wavy strip inserts. Exp Thermal Fluid Sci 50:107–113

Saisorn S, Wongwises S (2008) A review of two-phase gas-liquid adiabatic flow characteristics in micro-channels. Renew Sustain Energy Rev 12(3):824–838

Sasaki S, Kishimoto T (1986) Optimal structure for microgrooved cooling fin for high-power LSI devices. Electron Lett 22(25):1332–1334

Schilder B, Man SYC, Kasagi N, Hardt S, Stephan P (2010) Flow visualization and local measurement of forced convection heat transfer in a microtube. ASME J Heat Transf 132(3):031702

Schneider B, Kosar A, Kuo C-J, Mishra C, Cole GS, Scaringe RP, Peles Y (2006) Cavitation enhanced heat transfer in microchannels. J Heat Transfer Trans ASME 128(12):1293–1301

Schnurma R, Lardge M (1973) Enhanced heat flux in nonuniform electric fields. Proc R Soc London Ser A Math Phys Eng Sci 334(1596):71–82

Scortesse J, Manceau JF, Bastien F (2002) Displacement of droplets on a vibrating structure. Ultrasonics 40(1–8):349–353

Senthilraja S, Karthikeyan M, Gangadevi R (2010) Nanofluid applications in future automobiles: comprehensive review of existing data. Nano-Micro Lett 2:306–310
Serizawa A, Feng Z, Kawara Z (2002) Two-phase flow in microchannels. Exp Thermal Fluid Sci 26(6–7):703–714
Seth A, Lee L (1974) Effect of an electric field in presence of non-condensable gas on film condensation heat transfer. J Heat Transfer Trans ASME 96(2):257–258
Shah MM (1987) Improved general correlation for critical heat flux during upflow in uniformly heated vertical tubes. Int J Heat Fluid Flow 8(2):326–335
Sharp KV, Adrian RJ (2004) Transition from laminar to turbulent flow in liquid filled microtubes. Exp Fluids 36(5):741–747
Shirtcliffe NJ, McHale G, Atherton S, Newton MI (2010) An introduction to super hydrophobicity. Adv Colloid Interface Sci 161:124–138
Shoji Y, Sato K, Oliver DR (2003) Heat transfer enhancement in round tube using coiled wire: influence of length and segmentation. Heat Transfer Asian Res 32(2):99–107
Silk EA, Kim J, Kiger K (2006) Spray cooling of enhanced surfaces: impact of structured surface geometry and spray axis inclination. Int J Heat Mass Transf 49(25–26):4910–4920
Singh A, Ohadi M, Dessiatoun S (1997) EHD enhancement of in-tube condensation heat transfer to alternate refrigerant R-134a in smooth and microfin tubes, ASHRAE Trans Symp 813–823
Sminov G, Lunev V (1978) Heat transfer during condensation of vapour of dielectric field in electric field. Appl Electr Phenom USSR 2:37–42
Sobhan CB, Garimella SV (2001) A comparative analysis of studies on heat transfer and fluid flow in microchannels. Microscale Thermophys Eng 5(4):293–311
Somerscales EFC, Bergles AE (1997) Enhancement of heat transfer and fouling mitigation. Adv Heat Trans 30:197–253
Steinke ME, Kandlikar SG (2004) Single-phase heat transfer enhancement techniques in microchannel and minichannel flows. In: Proceedings of the second international conference on microchannels and minichannels (ICMM2004), Rochester, 17–19 June 2004. pp 141–148
Steinke ME, Kandlikar SG (2006a) Single-phase liquid friction factors in microchannels. Int J Therm Sci 45(11):1073–1083
Steinke ME, Kandlikar SG (2006) Single-phase liquid heat transfer in plain and enhanced microchannels. Paper no. ICNMM2006- 96227, ASME fourth international conference on nanochannels, microchannels and minichannels, 2006B, Limerick, 19–21 June. pp 943–951
Sun B, Huai X, Jiang DLRQ (2004) An investigation of enhancing natural convection of horizontal circular copper tube with acoustic cavitation. In: Proceedings of the 6th international symposium on heat transfer
Sunada K, Yabe A, Taketani T, Yoshizawa Y (1991) Experimental study of EHD pseudo-dropwise condensation. In: Proceedings of the ASME/JSME thermal engineering, vol 3
Tadrist L, Miscevic M, Rahli O, Topin F (2004) About the use of fibrous materials in compact heat exchangers. Exp Thermal Fluid Sci 28(2–3):193–199
Tardist L (2007) Review on two-phase instabilities in narrow spaces. Int J Heat Fluid Flow 28(1):54–62
Tauscher R, Mayinger F (1997) Enhancement of heat transfer in a plate heat exchanger by turbulence promoters. In: Proceedings of the international conference on compact heat exchangers for the process industries. Begell House, New York, pp 243–260
Taylor JB, Carrano AL, Kandlikar SG (2006) Characterization of the effect of surface roughness and texture on fluid flow—past, present, and future. Int J Therm Sci 45(10):962–968
Thakur R, Vial C, Nigam K, Nauman E, Djelveh G (2003) Static mixers in the process industries – a review. Chem Eng Res Des 81(A7):787–826
Thianpong C, Eiamsa-ard P, Wongcharee K, Eiamsa-ard S (2009) Compound heat transfer enhancement of a dimpled tube with a twisted tape swirl generator. Int Commun Heat Mass Transfer 36:698–704
Thianpong C, Eiamsa-ard P, Eiamsa-ard S (2012) Heat transfer and thermal performance characteristics of heat exchanger tube fitted with perforated twisted-tapes. Int J Heat Mass Transf 48:881–892

Thome JR, Dupont V, Jacobi AM (2004) Heat transfer model for evaporation in microchannels. Part I: presentation of the model. Int J Heat Mass Transf 47(14–16):3375–3385
Toda M (1979) Theory of air flow generation by a resonant type PVF2 bimorph cantilever vibrator. Ferroelectrics 22:911–918
Toda M (1981) Voltage-induced large amplitude bending device – PVF2 bimorph – its properties and applications. Ferroelectrics 32:127–133
Trisaksri V, Wongwises S (2007) Critical review of heat transfer characteristics of nanofluids. Renew Sustain Energy Rev 11(3):512–523
Trommelmans J, Berghmans J (1986) Influence of electric fields on condensation heat transfer of nonconducting fluids on horizontal tubes. In: Proceedings of the international heat transfer conference, vol 6. pp 2969–2974
Tsuzuki N, Kato Y, Nikitin K, Ishizuka T (2009) Advanced microchannel heat exchanger with S-shaped fins. J Nucl Sci Technol 46(5):403–412
Tuckerman DB (1984) Heat-transfer microstructures for integrated circuits. Stanford University, Stanford
Tuckerman DB, Pease RFW (1981) High-performance heat sinking for VLSI. IEEE Electron Device Lett ELD-2(5):126–129
Turlik I, Reisman A, Darveaux R, Hwang LT (1989) Multichip packaging for supercomputers. In: Proceedings of the technical program. NEPCON West '89, Anaheim, 13–15 June 1989. pp 37–58
Uhlenwinkel V, Meng R, Bauckhage K (2000) Investigation of heat transfer from circular cylinders in high power 10 kHz and 20 kHz acoustic resonant fields. Int J Therm Sci 39(8):771–779
Uttarwar SB, Raja Rao M (1985) Augmentation of laminar flow heat transfer in tubes by means of coiled wire inserts. Trans ASME 107:930–935
Velkoff H, Miller J (1965) Condensation of vapor on a vertical plate with a transverse electrostatic field. J Heat Transf 87(2):197–201
Walpole JN, Liau ZL, Diadiuk V, Missaggia LJ (1988) Microchannel heat sinks and microlens arrays for high average-power diode laser arrays. In: Proceedings of LEOS '88—lasers and electro-optics society annual meeting, Santa Clara, 2–4 Nov 1988. pp 447–448
Wang XQ, Mujumdar AS (2007) Heat transfer characteristics of nanofluids: a review. Int J Therm Sci 46(1):1–19
Wang BX, Peng XF (1994) Experimental investigation on liquid forced convection heat transfer through microchannels. Int J Heat Mass Transf 37(Suppl 1):73–82
Wang L, Sunden B (2002) Performance comparison of some tube inserts. Int Commun Heat Mass Transfer 29(1):45–56
Watson P (1961) Influence of an electric field upon heat transfer from a hot wire to an insulating liquid. Nature 189(476):563–564
Wawzyniak M, Seyed-Yagoobi J (1996) Experimental study of electrohydrodynamically augmented condensation heat transfer on a smooth and an enhanced tube. J Heat Transfer Trans ASME 118(2):499–502
Webb RL (1994a) Principles of enhanced heat transfer. Wiley, New York
Webb RL (1994) Advances in modeling enhanced heat transfer surfaces. In: Heat transfer 1994, proceedings of the 10th international heat transfer conference, vol 1. IChemE, Rugby, pp 445–459
Wei XJ, Joshi YK, Ligrani PM (2007) Numerical simulation of laminar flow and heat transfer inside a microchannel with one dimpled surface. J Electron Packag 129(1):63–70
Weigand B, Spring S (2011) Multiple jet impingement – a review. Heat Transfer Res 42(2, Sp Iss 5):101–142
Wilson OB (1988) Introduction to theory and design of sonar transducers. Peninsula, Los Altos
Wu HY, Cheng P (2003) An experimental study of convective heat transfer in silicon microchannels with different surface conditions. Int J Heat Mass Transf 46(14):2547–2556
Wu T, Ro P (2005) Heat transfer performance of a cooling system using vibrating piezoelectric beams. J Micromech Microeng 15(1):213–220

Xu B, Ooi KT, Wong NT, Choi WK (2000) Experimental investigation of flow friction for liquid flow in microchannels. Int Commun Heat Mass Transfer 27(Copyright 2001, IEE):1165–1176

Xu JL, Gan YH, Zhang DC, Li XH (2005) Microscale heat transfer enhancement using thermal boundary layer redeveloping concept. Int J Heat Mass Transf 48(9):1662–1674

Xu P, Yu B, Qiu S, Poh HJ, Mujumdar AS (2010) Turbulent impinging jet heat transfer enhancement due to intermittent pulsation. Int J Therm Sci 49(7):1247–1252

Xuan YM, Li Q (2000) Heat transfer enhancement of nanofluids. Int J Heat Fluid Flow 21(1):58–64

Yabe A, Kikuchi K, Taketani T, Mori Y, Hijikata K (1982) Augmentation of condensation heat transfer by applying non-uniform electric field. In: Proceedings of the seventh international heat transfer conference. pp 189–194

Yabe A, Taketani T, Kikuchi K, Mori Y, Miki H (1982) Augmentation of condensation heat transfer by applying electro-hydro-dynamical pseudodrop wise condensation, pp 2957–2962

Yabe A, Taketani T, Maki H, Takahashi K, Nakadai Y (1992) Experimental study of electrohydrodynamically (EHD) enhanced evaporator for nonazeotropic mixtures. ASHRAE Trans 98:455–461

Yamashita K, Yabe A (1997) Electrohydrodynamic enhancement of falling film evaporation heat transfer and its long-term effect on heat exchangers. J Heat Transfer Trans ASME 119(2):339–347

Yamashita K, Kumagai M, Sekita S, Taketani T, Kikuchi K (1991) Heat transfer characteristics on EHD condenser. Proc ASME/JSME Therm Eng 3:61–67

Yang S (2002) Numerical investigation of heat transfer enhancement for electronic devices using an oscillating vortex generator. Numer Heat Transfer Part A Appl 42(3):269–284

Yang S, Zhang L, Xu H (2011) Experimental study on convective heat transfer and flow resistance characteristics of water flow in twisted elliptical tubes. Appl Therm Eng 31:2981–2991

Yu H, Kwon JW, Kim ES (2006) Microfluidic mixer and transporter based on PZT self-focusing acoustic transducers. J Microelectromech Syst 15(4):1015–1024

Yu WH, France DM, Routbort JL, Choi SUS (2008) Review and comparison of nanofluid thermal conductivity and heat transfer enhancements. Heat Transfer Eng 29(5):432–460

Zhang L, Koo J-M, Jiang L, Asheghi M, Goodson KE, Santiago JG, Kenny TW (2002) Measurements and modeling of two-phase flow in microchannels with nearly constant heat flux boundary conditions. J Microelectromech Syst 11(1):12–19

Zhang LK, Goodson KE, Kenny TW (2004) Silicon microchannel heat sinks: theories and phenomena. Springer, Berlin

Zhang HY, Pinjala D, Iyer MK, Nagarajan R, Joshi YK, Wong TN, Toh KC (2005) Assessments of single-phase liquid cooling enhancement techniques for microelectronics systems. In: Proceedings of ASME/Pacific Rim technical conference and exhibition on integration and packaging of MEMS, NEMS, and electronic systems: advances in electronic packaging 2005, PART A, San Francisco, 17–22 July 2005. pp 43–50

Zhang T, Tong T, Chang J-Y, Peles Y, Prasher R, Jensen M, Wen J, Phelan P (2009) Ledinegg instability in microchannels. Int J Heat Mass Transf 52(25–26):5661–5674

Zhou JW, Wang YG, Middelberg G, Herwig H (2009) Unsteady jet impingement: heat transfer on smooth and non-smooth surfaces. Int Commun Heat Mass Transfer 36(2):103–110

Part II
Enhancement of Heat Transfer in Two-Phase Flow

Abstract Enhancement of heat transfer in two-phase flow is an evergreen and important topic of huge relevance in existing, future and renewable energy systems as well as for energy conservation and environment protection. This part presents a state-of-the-art overview of heat transfer enhancement techniques for two-phase flow and mainly highlights those either commercially used techniques or the most recent emerging methods. Performances and characteristics of passive enhancement techniques (e.g., surface coating, roughened and finned surfaces, insert devices, curved geometries and additives) and active enhancement techniques (e.g., electrohydrodynamic phenomenon, jet impingement) are briefly described. Several recent enhancement techniques, e.g., nanoscale surface coatings, microfin tubes and nanoparticle additives, are outlined for their promising potential in enhancing phase-change heat transfer. As boiling and condensation are complex phenomena, heat transfer enhancement in two-phase flow calls for continued research and development. Several recommendations are given for future work.

Keywords Heat transfer enhancement • Boiling • Condensation • Two-phase flow • Surface coating • Fin • Nanofluid • Electrohydrodynamics

Chapter 8
Introduction

Heat transfer enhancement is an evergreen and important topic of huge relevance in existing, future and renewable energy systems as well as for energy conservation and environment protection (Sundén and Wu 2015). Enhancement is normally concerned with increasing the heat transfer coefficient. The goal of enhancement techniques might be to reduce the size of the heat exchanger for a given duty, to increase the capacity of an existing heat exchanger, or to reduce the approach temperature difference (Reay et al. 2013). Due to advancement in manufacturing processes, enhanced surfaces/geometries have been routinely used in nearly all heat exchangers in the air-conditioning and automotive industries, and gradually implemented in the electronics cooling, process, and power industries (Webb and Kim 2005).

Heat transfer enhancement techniques can be classified either as passive (no external power needed) or as active (external powered required). Passive techniques employ special surface geometries (e.g., surface coatings, fins, inserts, curved tubing, surface tension devices), or fluid additives (e.g., microparticles, nanoparticles, phase change materials, surfactants) for heat transfer enhancement. Bergles (2002) defined the four generations of heat transfer technology using passive techniques. Most of the heat transfer enhancement techniques were covered in the book, i.e., Principles of Enhanced Heat Transfer, by Webb and Kim (2005).

Two-phase flow involves any two of the three phases (i.e., solid, liquid and gas), which is common in a wide variety of natural phenomena and technical processes, such as dust storms, flash floods, inkjets, deposition and coating, combustion, boiling, condensation and frost etc. Phase change two-phase flow (e.g., boiling, condensation) is a broad field that finds applications in many engineering disciplines as they are generally associated with high heat transfer rates. Among numerous challenges to meet the rising global energy demand and to mitigate the corresponding cooling bottlenecks, enhancing phase change heat transfer has received much attention for decades. This chapter will focus on the recent development of heat transfer enhancement techniques for two-phase flow, mainly for boiling and condensation. Boiling and condensation processes are compared, as shown in Fig. 8.1, according to their physical processes, time scales and length scales. Boiling and condensation

S.K. Saha et al., *Advances in Heat Transfer Enhancement*,
DOI 10.1007/978-3-319-29480-3_8

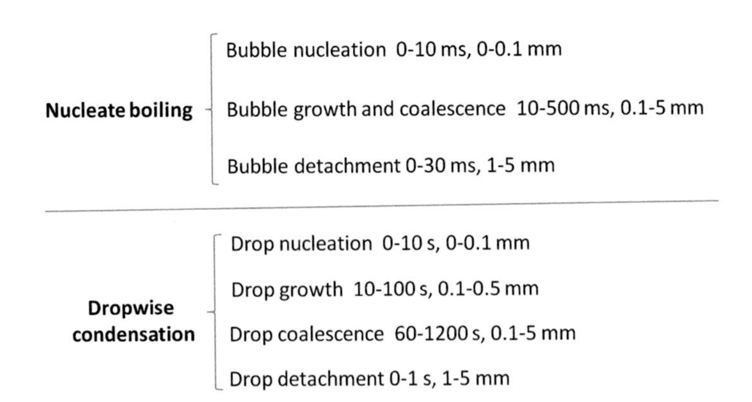

Fig. 8.1 A representation of the time and length scales of nucleate boiling and dropwise condensation

are complicated processes as these cover a wide range of length scales and time scales. Phase change heat transfer covers length scales and time scales of several orders of magnitude, from those relevant to nucleation (O(nm) and O(ms)) to those associated with flow instabilities (O(mm) and O(minute)).

Any fluid motion is controlled by forces. Basically, there are three categories of forces: volume forces (body forces), surface forces and line forces. Five or six forces come into play for gas-liquid two-phase flow (Wörner 2003; Kandlikar 2010; Wu et al. 2013a), such as pressure force, gravity (or buoyancy), inertia, viscous force, surface tension and evaporation momentum force for boiling. Table 8.1 summarizes the major forces and their magnitudes in gas-liquid two-phase flow. Parameter L is the characteristic length. The evaporation momentum force, as described in Kandlikar (2010), acts on the evaporating interface and plays a major role in its motion. The evaporation momentum is not present in non-boiling two-phase flows. As shown in Table 8.1, these forces have different dependences on the characteristic length. Therefore, the relative magnitudes of these forces may vary greatly with the characteristic length. For example, the surface tension force, which might be comparably negligible for large elements, could play an important role for small elements such as microfins. Heat transfer enhancement can be achieved by properly modifying the relative magnitude of those forces locally or globally. For example, surface tension can be varied either by changing the local curvature or by modifying the surface wettability. Corrugations affect the near-wall velocity profiles and thus affect the local inertia force and viscous force. Apart from the forces in Table 8.1, additional forces can be introduced to enhance heat transfer, such as electrohydrodynamic (EHD) forces or surface vibration.

The mechanism of heat transfer enhancement is a strong function of the nature of the fluid stream (gas, liquid, liquid-gas or liquid-vapor mixture, in some cases solid, or a mixture of liquid, gas and solid) and the mode of heat transfer (e.g., natural convection, forced convection, boiling, condensation, radiation). For single-phase flow, extended fins are commonly used to enhance heat transfer by increasing the heat transfer area and modifying the velocity profile (to produce turbulence).

Table 8.1 Major forces and their magnitude in two-phase flow

		Magnitude of		
Forces	Type	Force	Force per unit area	Force per unit volume
Pressure force	Surface force	$A\Delta p$	Δp	$\Delta p/L$
Gravity, buoyancy	Volume force	$Vg\rho$, $Vg\Delta\rho$	$Lg\rho$, $Lg\Delta\rho$	$g\rho$, $g\Delta\rho$
Inertia	Volume force	$V\rho U^2/L$	ρU^2	$\rho U^2/L$
Viscous force	Surface force	$A\mu U/L$	$\mu U/L$	$\mu U/L^2$
Surface tension	Line force	$L\sigma$	σ/L	σ/L^2
Evaporation momentum	Surface force	$A(q/h_{lv})^2/\rho_v$	$(q/h_{lv})^2/\rho_v$	$(q/h_{lv})^2/(\rho_v L)$

For phase change two-phase flow, surface engineering techniques which modify the heating/cooling surface parameters (e.g., surface roughness, wettability, porosity) are of increasing interest. Surface modification techniques such as surface machining and roughening, coating, wettability engineering, etc., can enhance boiling and condensation heat transfer by increasing nucleation sites, modifying the bubble/droplet dynamics, modifying the flow patterns, improving capillary wetting for boiling to induce thin film evaporation, or sustaining dropwise condensation and ensuring adequate condensate drainage. Flow patterns are closely related to heat transfer for two-phase flow. For flow boiling, there are typical flow patterns such as bubbly flow, slug flow, plug flow, churn flow, annular flow, and stratified flow for horizontal and inclined orientations. As heat transfer mechanisms might be different in different flow patterns, enhancement techniques need to be selected with special care for different flow patterns. For example, in bubbly flow, the heat transfer mechanism is nucleate boiling. Heat transfer enhancement can be achieved by providing more stable nucleation sites, decreasing the bubble departure diameter and increasing the bubble departure frequency. For annular flow, as there is little bubble nucleation in the thin liquid films between the vapor core and the wall, convective boiling is the dominant heat transfer mechanism. Boiling heat transfer in annular flow might be enhanced by thinning the liquid films while delaying the liquid film dryout. The two heat transfer modes for condensation are filmwise condensation and dropwise condensation. The condensation heat transfer coefficient of dropwise condensation is orders of magnitude higher than that of filmwise condensation. Therefore, it is very important to enhance droplet mobility and condensate drainage to sustain dropwise condensation and prevent formation of liquid film on the surface. A very common method to enhance filmwise condensation is to use surface tension devices such as flutes to facilitate condensate removal in concave portions by shear force or gravity and to maintain thin liquid films on the convex portions with high condensation rates. Table 8.2 shows the various forces which affect bubble/droplet dynamics, flow patterns and corresponding heat transfer mechanisms for boiling and condensation.

Some surface modification techniques might pose characteristics both favorable and unfavorable for heat transfer enhancement. For example, hydrophobic surfaces may introduce more nucleation sites which enhance heat transfer, while the nucleated bubbles may coalesce easily to form a vapor layer on the heating surface to

Table 8.2 List of forces, flow patterns and dominant heat transfer mechanisms for boiling and condensation

	Driving forces promoting discrete phase departure	Restraining forces impeding discrete phase departure	Flow patterns	Dominant heat transfer mechanisms
Boiling	Buoyancy, shear, inertia	Surface tension	Bubbly flow	Nucleate boiling
			Intermittent flow	Both
			Annular flow	Convective boiling
Condensation	Gravity, surface tension, shear	Surface tension, gravity, inertia	Annular flow	Filmwise condensation
			Intermittent flow	Both
			Bubbly flow	Single phase convection

cause boiling crisis and severe heat transfer deterioration. Porous coatings provide more stable nucleation sites and capillary wetting which are favorable for heat transfer, but at the same time the coatings pose a thermal resistance to heat transfer.

While heat transfer enhancement of boiling and condensation is the topic of several books and book chapters (e.g., Thome 1990; Rohsenow 1998; Webb and Kim 2005; Sundén and Wu 2015) and several review papers (e.g., Marto 1986; Bergles 1997; Rose 2004; Webb 2004), a state-of-the-art overview of heat transfer enhancement for two-phase flow (mainly boiling and condensation) is still missing. This article aims to close this gap and will mainly focus on those either commercially used or recently developed and promising enhancement techniques.

Chapter 9
Passive Techniques

Passive techniques are commonly used for heat transfer enhancement as no external energy is needed. Many passive enhancement techniques were given in details by Webb and Kim (2005). However, recently developed enhancement techniques such as nanoscale structures and composite micro/nanostructures are not mentioned. In general, this part will mainly cover surface coatings (microscale and nanoscale), roughened or finned surfaces, inserts, curved geometries, surface tension devices and additives (e.g., nanoparticles, surfactants).

9.1 Surface Coatings

9.1.1 Macro/Microporous Coatings

Porous coatings containing re-entrant-type cavity geometries can enhance boiling heat transfer (Webb 1983; Chang and You 1997; Webb and Kim 2005). Many commercial enhanced surfaces have been shown to provide these geometries, such as ECR-40 of Furukawa Electric, GEWA series of Wieland-Werke, High-Flux of UOP, Thermoexcel series of Hitachi and Turbo-B of Wolverine (Ammerman and You 2001). Ammerman and You (2001) experimentally investigated the convective boiling performance of FC-87 in a horizontal small channel with and without, a microporous surface coating. The coated channel shows substantial heat transfer and critical heat flux (CHF) enhancement over the uncoated channel, especially for low subcooling cases. Boiling occurs at much lower wall superheats for the coated channel. The presence of the coating only has marginal effects on pressure drop for subcoolings of 15 and 24 °C. The presence of the additional nucleation sites due to the porous coating will promote boiling. At the same time, the vapor generated from these sites is quickly condensed at high subcooling levels, and therefore only causes little pressure drop penalty. However, the pressure drop is higher for the coated surface than the uncoated

S.K. Saha et al., *Advances in Heat Transfer Enhancement*,
DOI 10.1007/978-3-319-29480-3_9

surface at a low subcooling of 3 °C. Based on the qualitative comparison of heat transfer and pressure drop, the coated surface provides an advantage over the uncoated surface. Porous coatings can enhance boiling heat transfer with initiation of boiling at lower wall superheats, increased heat transfer coefficients, and elevated CHF levels. The mechanism of heat transfer enhancement attributes to the increase in number of nucleation sites and bubble departure frequency, while the improvement in flow stability and capillary wetting enhance CHF levels.

Dawidowicz and Cieśliński (2012) investigated the heat transfer and pressure drop during flow boiling of pure refrigerant and refrigerant/oil mixtures in tubes coated with metallic porous layers against the inner wall. The thickness, porosity and mean pore radius of the porous coating are 55 μm, 18 % and 1.45 μm, respectively. For pure refrigerants R22, R134a and R407C, the porous coating results in a higher (about five to six times, as shown in Fig. 9.1a) average heat transfer coefficient and simultaneously in a lower pressure drop compared with a smooth stainless steel tube. Heat transfer deterioration for refrigerant/oil mixtures was observed and the average heat transfer coefficient degradation was more distinct in tubes with porous coating. Oil is present only in the liquid phase. Generally, oil has higher viscosity and surface tension than refrigerants. As the oil concentration increases along the flow direction in the liquid phase, the higher surface tension impedes bubble formation and increases tube wetting, while higher viscosity reduces the velocity of the liquid phase. Thus, both nucleate boiling and convective boiling tend to be degraded by oil-rich liquid sub-layers. As shown in Fig. 9.1b, the absolute value of the average heat transfer coefficient for the refrigerant/oil mixture (1 % of oil concentration) is still about two to three times higher than that in a smooth tube.

Porous coatings including sintered microparticles can provide more thin-film area among the microparticles in annular flow at relatively large vapor qualities to enhance thin film evaporation. A thin-film region consists of three regions: the nonevaporating thin film region, the evaporating thin film region, and the meniscus thin film region (Hanlon and Ma 2003). In the presence of a thin film, a majority of the heat will be transferred through a very small region. Maximum evaporation and heat transfer occur from the evaporating thin film region and the liquid is fed from the bulk liquid

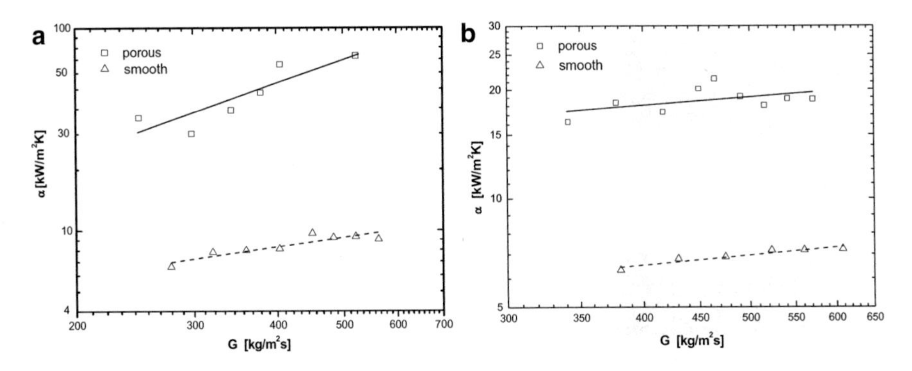

Fig. 9.1 Average heat transfer coefficient for flow boiling of (**a**) pure refrigerant R22, and (**b**) R134a/oil mixture with 1 % oil concentration in a smooth stainless steel tube and in a tube with porous coating. Dawidowicz and Cieśliński (2012)

through the intrinsic meniscus region. Another important feature of porous coating is the CHF enhancement due to capillary wetting. Recently, modulated porous coatings were proposed to enhance both heat transfer coefficient and CHF by designing separate liquid and vapor pathways. The liquid and vapor flow resistances can be decreased due to these separate and controlled pathways. Figure 9.2 shows micrographs of modulated porous-layer coatings proposed by Liter and Kaviany (2001). The coatings are made from spherical copper particles of diameter 200 μm. During sintering at 1015 °C, the desired modulation was shaped using open-faced graphic molds. The resulting fabricated modulations consisted of either conical stacks of particles or of particle walls in a waffle pattern arranged in either square or hexagonal arrays across the surface. The modulation coatings create alternating regions of low resistance to vapor escape and high capillary-assisted liquid feeding. This would result in preferential liquid-vapor counterflow paths within the porous coatings to facilitate heat transfer from the surface to the liquid pool in a manner similar to that of heat pipes. Qu et al. (2012) cross-cutted the metal foam, as shown in Fig. 9.3, and therefore provides more spaces for vapor escape. The grooved metallic foam surface can separate the escaping vapor bubbles from the sucked liquid in the porous structure. At the same time, sufficient liquid can be supplied by capillary force for bubble nucleation. The heat transfer coefficient and CHF are enhanced simultaneously. However, all these works on modulated porous coatings were conducted for pool boiling. More investigations on flow boiling heat transfer enhancement need to be performed in the future.

9.1.2 *Nanoscale Surface Coating*

Due to recent advancement in nanotechnology, nanostructured coatings are being developed to enhance phase-change heat transfer. Generally, the surface wettability is modified to affect the heat transfer performance (Wang and Dhir 1993). The nanostructured coatings can be either nano-porous or composed of nano-fins. A

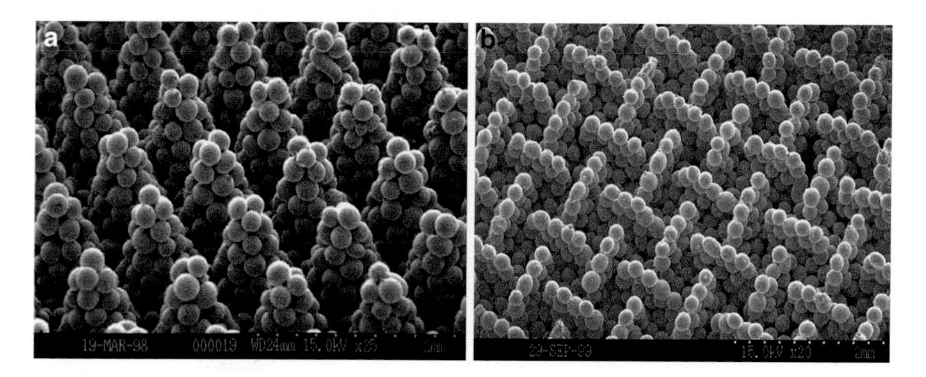

Fig. 9.2 The modulated porous coating contains spherical copper particles of diameter $d=200$ μm molded into (**a**) conical particle stacks, and (**b**) particle walls in a waffle pattern arranged in a square array. Adapted from Liter and Kaviany (2001)

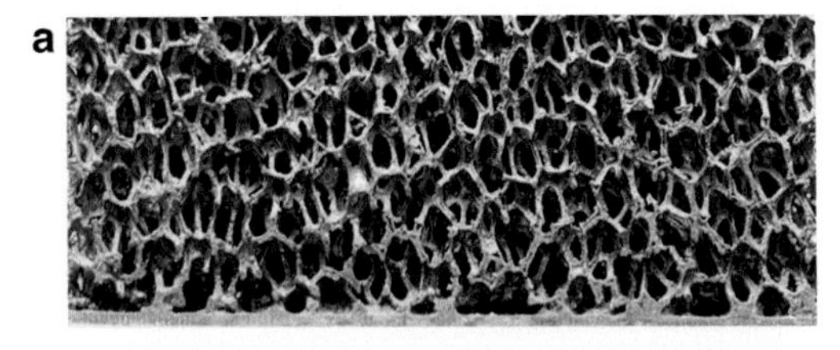

Fig. 9.3 Metal foam sample: (**a**) photograph of copper foam with copper substrate and (**b**) photograph of V-shaped grove array. Qu et al. (2012)

nanoporous coating augments the boiling heat transfer rate due to the water entrapping capability of pores (Sarwar et al. 2007). The trapped water will flow out from the pores due to the local pumping action during heating, which leads to an increase in nucleation sites for heat transfer enhancement. Nano-fins like carbon nanotubes (CNTs) increase the two-phase heat transfer due to the scale effect (increased surface area or "nano-fin" effect) (Khanikar et al. 2009; Singh et al. 2010) and high thermal conductivity (as high as 3000 W/(mK)). The increased surface area by CNT coating leads to effective vapor embryo entrapment (Ujereh et al. 2007), which initiates and enhances the nucleation for boiling heat transfer. Kumar et al. (2014) used diamond as an intermediate layer to coat CNTs on a sand-blasted copper substrate. The diamond intermediate layer can enhance the adhesion between the CNTs and copper and thus reduce the interfacial resistance. The surface morphology of the sand blasted copper substrate, the diamond coating, and the CNT coating are shown in Fig. 9.4. The results show that CNT coatings enhance the boiling heat transfer due to higher nucleation density and high thermal conductivity of CNT. Enhancement in CHF was observed and the enhancement is higher at lower mass fluxes. Under fierce flow boiling conditions, the vertically aligned CNTs might bend upon the surface at relatively large mass fluxes and thus the surface morphology changes with time. Stability analysis of CNT coatings with respect to time was also conducted. As shown in Fig. 9.5, there is no obvious difference at a low mass flux of 95 kg/m^{2}s. However, an appreciable change in boiling curves was observed for CNT coatings at a relatively large mass flux of 283 kg/m^{2}s, which indicates morphology change of the CNT coatings with respect to time. The gradual deterioration of CHF enhancement for the higher mass flux is due to the folding of CNTs towards the surface, which in turn reduces the size of void formed between the two CNTs. But at lower mass flux this bending of CNTs takes place slowly, so no appreciate change in critical heat flux was found after four tests.

Kousalya et al. (2012) proposed that flow boiling heat transfer from a CNT-coated copper surface can be further enhanced with light exposure. Well-anchored CNTs were grown on roughened copper substrates using a microwave plasma chemical vapor deposition technique. The diameter of the CNTs varied roughly between 30 and 80 nm. Field emission scanning electron microscopy (FESEM)

Fig. 9.4 SEM images of (**a**) sand blasted copper, (**b**) diamond coated copper, and (**c**) CNT-coated copper substrates. Kumar et al. (2014)

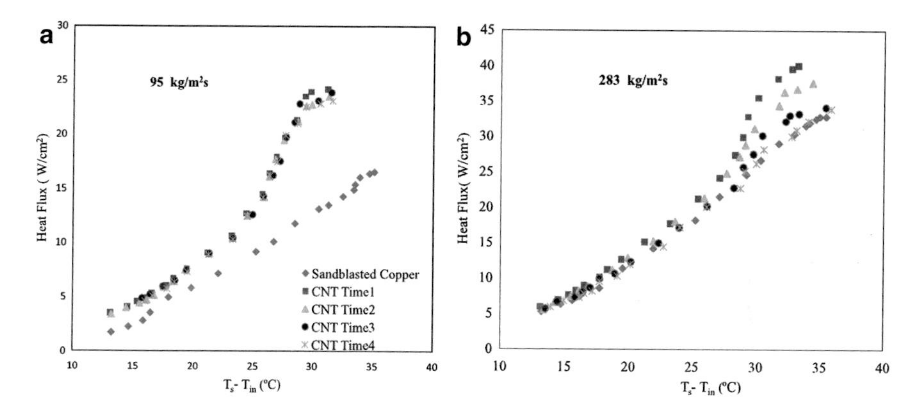

Fig. 9.5 Subcooled flow boiling curves measured at (**a**) 95 kg/m^{2}s and (**b**) 283 kg/m^{2}s. Kumar et al. (2014)

images (Fig. 9.6) revealed that the CNTs formed a moderately dense and randomly oriented mesh on the surface of the copper substrate. A total of nine flow boiling tests were conducted (T1-T9), two tests with the bare copper substrate (T1, T2) and seven tests with the CNT-coated copper substrate (T3-T9). Figure 9.7a compares the flow boiling performance with and without UV-visible light for the bare copper substrate. No significant enhancement was detected due to UV-visible illumination. Figure 9.7b depicts the results obtained for the seven boiling tests with the CNT-coated substrate. The thermal characteristics of the CNT-coated substrate with light exposure (T4, T6, and T8) are consistent and within experimental uncertainties. As shown, there is a pronounced effect due to photonic excitation of the CNT-coated surface compared to the behavior prior to the first illumination sequence (T3). It was calculated that the average onset of nucleate boiling was reduced by 4.6 °C and the effective heat transfer coefficient increased by 41.5 %. Interestingly, for the experiments conducted after darkness conditioning, the boiling curves do not revert to that of the original non-illuminated experiment. If the darkness-conditioned boiling curves (T5, T7, and T9) are considered separately, a leftward shifting trend is observed, suggesting an increased retention of the surface change that produced the original enhancement with increase in total time duration of photonic excitation. All tests were conducted at a very low mass flux. It is supposed that the heat transfer enhancement will probably degrade at high mass fluxes.

Fig. 9.6 FESEM image of CNTs grown on copper substrate: (**a**) low magnification and (**b**) high magnification-prior to testing; (**c**) High magnification-post testing and (**d**) flower like CuO nanostructures. Kousalya et al. (2012)

Filmwise condensation is quite common in industrial components. The formation of a liquid film is not desired due to the large resistance to heat transfer (Fig. 9.8a). Methods have been proposed to induce dropwise condensation, such as coating with a low-energy non-wetting "promoter" material (i.e., polymer coating, wax, long-chain fatty acid). Recent advancement in nanofabrication has allowed for the development of superhydrophobic surfaces, on which nearly spherical water droplets form with high mobility and minimal droplet adhesion. In addition, the role of surface structuring on wetting characteristics have been studied in detail (Patankar 2010; Bocquet and Lauga 2011; Kang et al. 2011) to enhance condensation heat transfer performance. Under Earth gravity conditions, the droplet size has to approach the capillary length ($[\sigma/(\rho g)]^{0.5}$, $\approx$2.7 mm) to overcome the contact line pinning force to depart from the surface (Fig. 9.8b). Recently, some researchers (Boreyko and Chen 2009; Narhe et al. 2009; Chen et al. 2011; Rykaczewski and Scott 2011; Liu et al. 2012; Miljkovic and Wang 2013; Miljkovic et al. 2013) found that when small water droplets ($\approx$10–100 μm) merge on suitably designed superhydrophobic surfaces, the coalesced droplets jump out of the surface due to the release of excess surface energy (Fig. 9.8c). Jumping droplet condensation has offered a new avenue to further enhance heat transfer by increasing the time-averaged density of small droplets (Miljkovic and Wang 2013). However, if the nucleation density is too high, droplet jumping cannot be sustained. Discrete non-jumping droplets form and adhere strongly to the surface, leading to a flooding condensation mode with worse heat transfer performance (Fig. 9.8d). Figure 9.8e shows heat transfer measurements of steam condensation at near-atmospheric pressure. The superhydrophobic surface with jumping droplets achieves the best heat transfer performance.

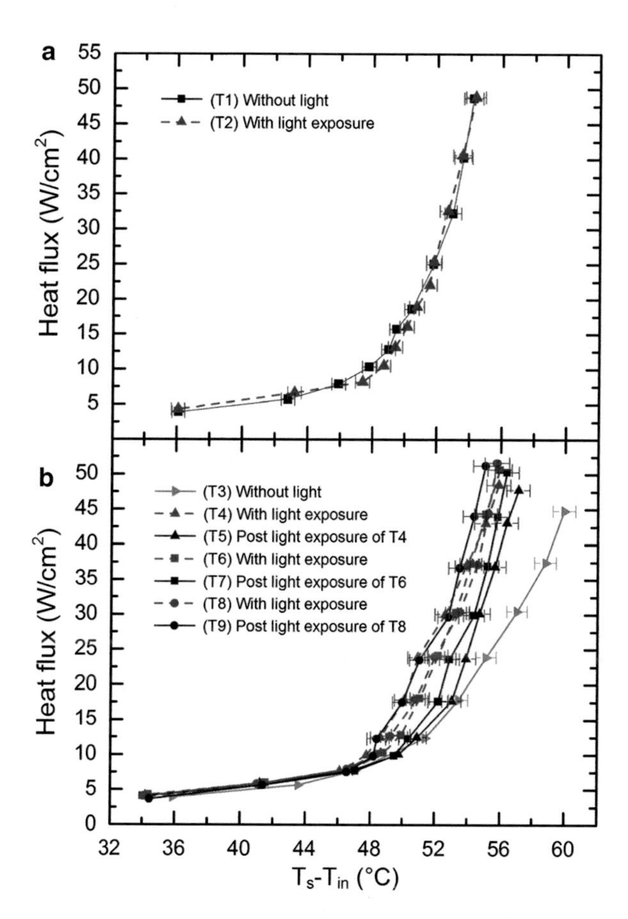

Fig. 9.7 Flow boiling curves for (**a**) bare copper surface and (**b**) CNT-coated copper surface with $G \approx 38$ kg/m^{2}s and $T_{in} = 59.5 \pm 0.5$ °C. Kousalya et al. (2012)

Cheng et al. (2012) investigated water vapor condensation on a two-tier superhydrophobic surface in an environmental scanning electron microscope. Figure 9.9 shows the two-tier structures consisting of CNTs on micromachined pillars. Both dropwise condensation and filmwise condensation were observed. Especially, a thick condensate film was captured on the textured surface and the condensate cannot be automatically removed from the textured condensing surface. It seems that condensate flooding becomes more pronounced with the increase of the heat flux and the saturation temperature. As a result, the condensation heat transfer coefficient on the two-tier surface is worse than that on a smooth hydrophobic surface and even not as efficient as a bare silicon surface, as shown in Fig. 9.10. Therefore, superhydrophobic surfaces cannot always improve heat transfer. Proper surface structuring is the key to induce dropwise condensation and prevent condensate flooding.

The movements of droplets can also be induced by unbalanced surface tension forces, known as the Marangoni effect. In order to surmount the contact line pinning force, additional energy must be supplied to the droplet by using the force arising from the surface-tension gradient to bias the droplet motion. Figure 9.11 shows a

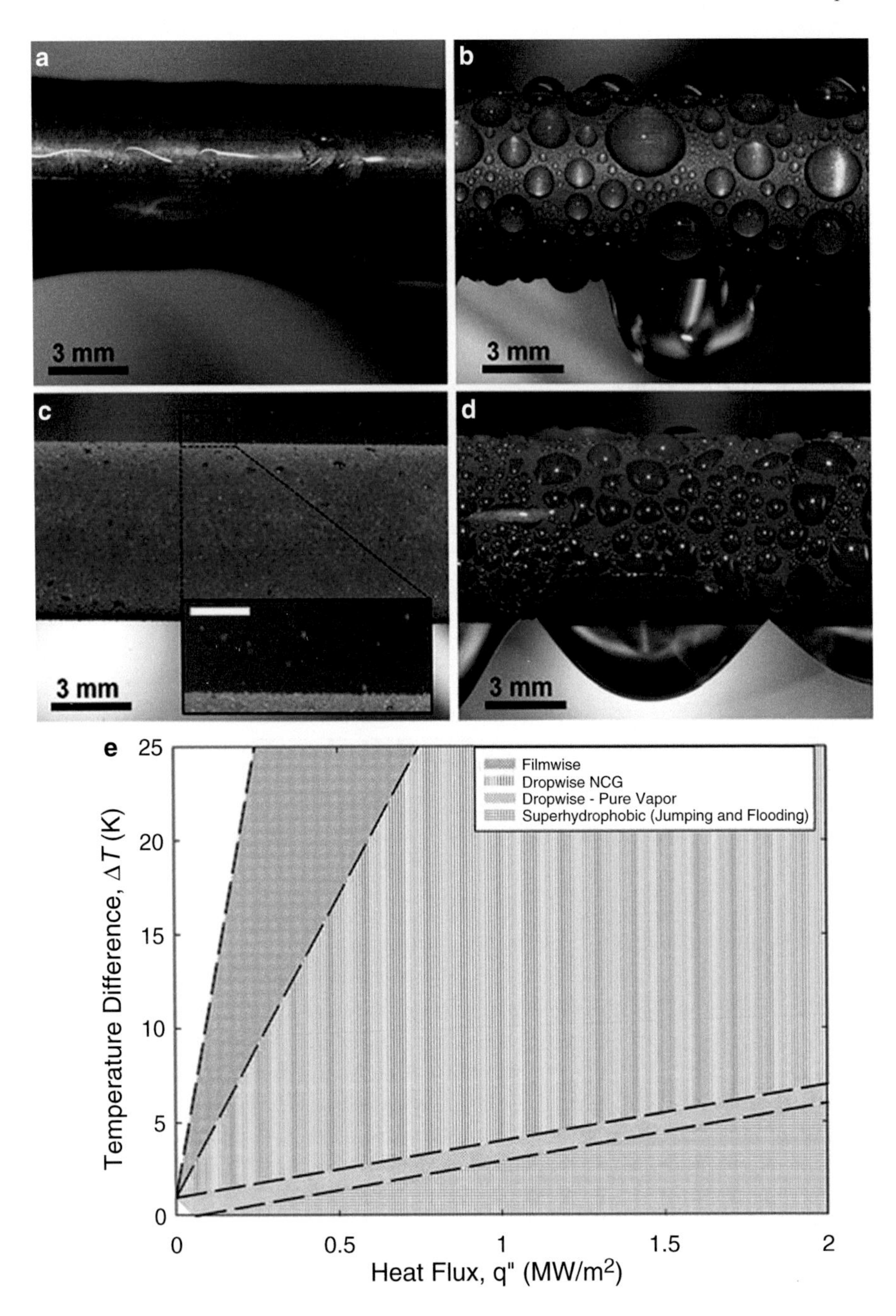

Fig. 9.8 Condensation heat transfer modes and performance. Images of (**a**) filmwise condensation on a smooth hydrophilic Cu tube, (**b**) dropwise condensation on a silane coated smooth Cu tube, (**c**) jumping-droplet superhydrophobic condensation on a nanostructured CuO tube (*inset*: magnified view of the jumping phenomena, scale bar is 500 μm), (**d**) flooding condensation on a nanostructured CuO tube, (**e**) heat transfer measurements of steam condensation at near-atmospheric pressure ($\Delta T = T_{sat}(P_v) - T_{wall}$). Note: *NCG* non-condensable gas. Miljkovic and Wang (2013)

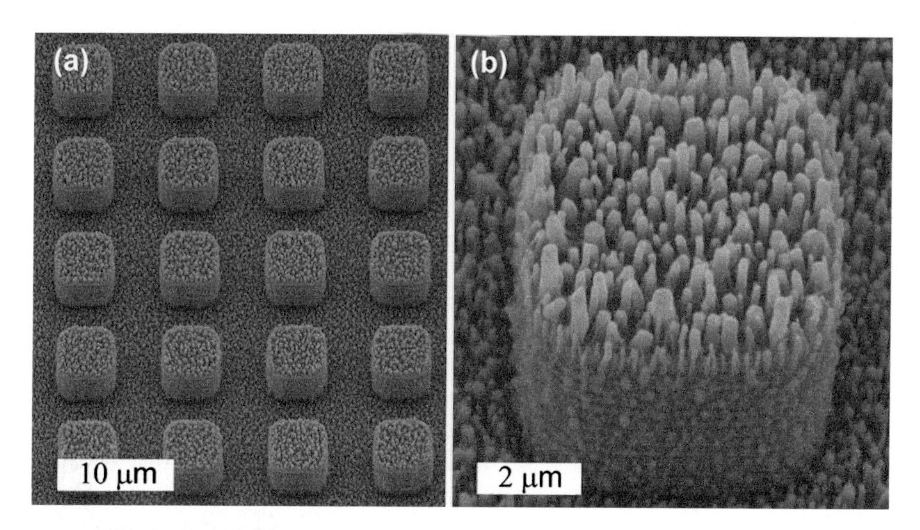

Fig. 9.9 Two-tier structures consisting of CNTs on micromachined pillars. (**a**) overall structure, (**b**) detailed structure. Cheng et al. (2012)

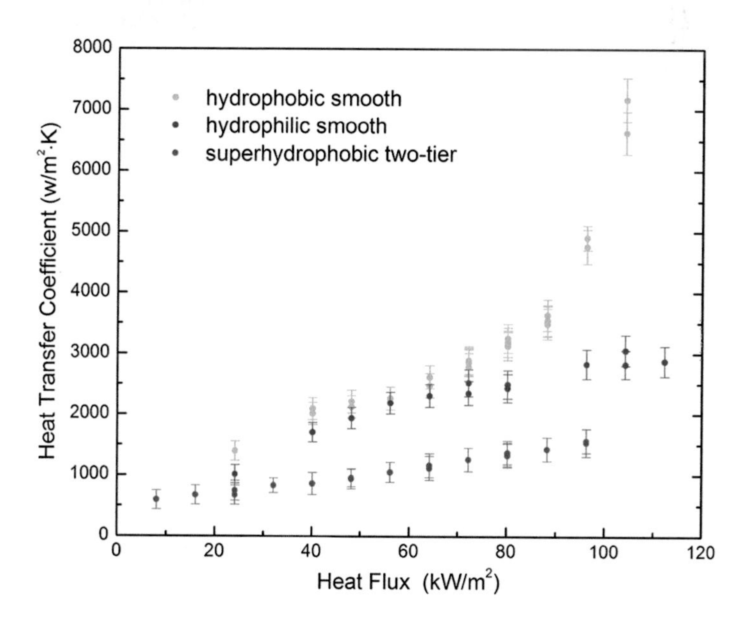

Fig. 9.10 Comparison of heat transfer coefficients on upside-down surfaces. Cheng et al. (2012)

Fig. 9.11 Schematic of droplet motion on a wettability gradient surface

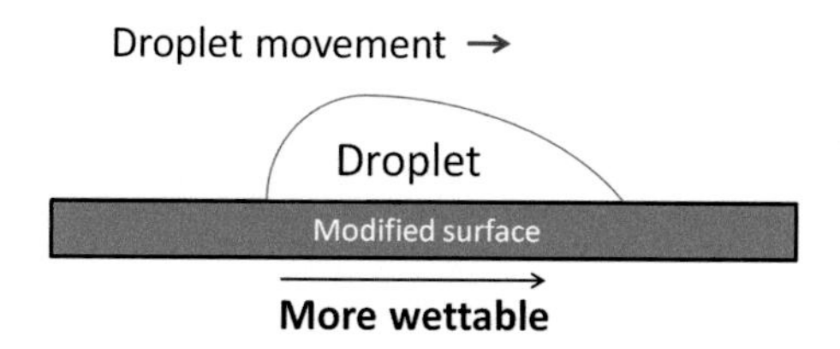

wettability gradient surface. The droplet will move from the less wettable towards the more wettable direction. Daniel et al. (2001) reported a new type of surface tension-guided flow, in which the drops move hundreds to thousands of times faster than the speeds of typical Marangoni flows. The test surface had a radially outward gradient of chemical composition that was prepared by diffusion-controlled silanization. The central part of the silicon surface, closest to the drop, became maximally hydrophobic, whereas its peripheral zone remained wettable by water with a near-zero contact angle. In air, small drops (1–2 mm) of water move on such a surface radially outward with typical speeds of 0.2–0.3 cm/s. However, when saturated steam (100 °C) condenses on such a surface, even smaller drops (0.1–0.3 mm) attain speeds that are two orders of magnitude higher than those observed under ambient conditions. Comparing with filmwise condensation on a horizontal unmodified surface, dropwise condensation on the gradient surface enhances the heat transfer by a factor of three when the surface is subcooled (ΔT) from that of steam by about 20 °C and by a factor of 10 when subcooling is about 2 °C.

9.2 Roughened and Finned Surfaces

Roughened and finned surfaces are very efficient for phase-change heat transfer enhancement, such as the four patented structured boiling surfaces shown in Webb (2004) and fluted surfaces for condensation (Fig. 9.12). There are two main heat transfer mechanisms for flow boiling in convectional channels: nucleate boiling and convective evaporation. Generally, structured surfaces are very efficient for nucleate boiling as these structures can provide more stable nucleation sites than smooth surfaces. However, the structured surfaces might not be sufficient for convective boiling. Enhanced boiling tubes with these structures generally yield a convective evaporation component typical of that of a plain tube (Webb and Kim 2005).

Fluted surface is a type of surface tension device which can enhance film condensation heat transfer. Gregorig (1954) was the first to propose that surface tension can play an important role during the condensation process. If the convex fin profile is of the proper shape, Gregorig showed that surface tension force would drain the

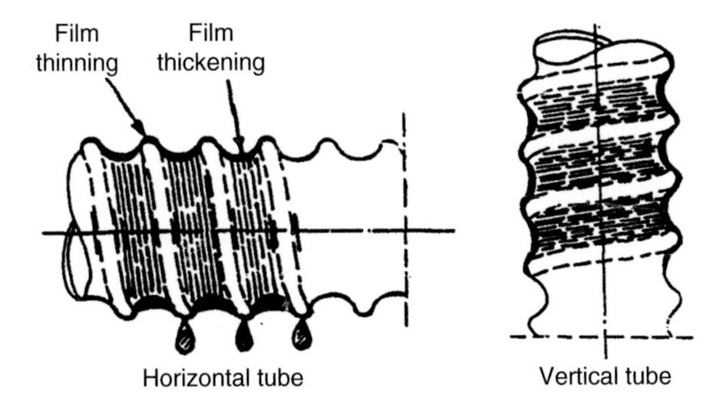

Fig. 9.12 Horizontal and vertical spirally fluted tubes. Aly and Bedrose (1995)

condensate from the convex part to the concave part. Then, the accumulated condensate in the concave part is drained by gravity and shear force. Similarly, fins can also induce condensate drainage (Wanniarachchi et al. 1986).

Enhanced tube geometries used for tube-side convective vaporization generally promote the convective evaporation term, rather than nucleate boiling. As the third generation heat transfer technology, microfin tube belongs to the most common passive heat transfer enhancement devices in many applications including compact heat exchangers, refrigeration and air-conditioning systems, petrochemical processes, waste heat recovery etc., due to its large heat transfer enhancement and moderate pressure drop penalty (Wu et al. 2015; Wu and Sundén 2015a). The microfin internal geometry can provide a decent heat transfer enhancement with a relatively low pressure drop penalty (Liebenberg and Meyer 2007; Li et al. 2012). Heat transfer characteristics of R410A evaporation in one plain (smooth) tube and five microfin tubes are presented in Fig. 9.13 (Wu et al. 2013b). Geometries of the microfin tubes are shown in Table 9.1. The heat transfer coefficient of R410A evaporation inside microfin tubes is about 1.70–2.85 times that of the plain tube. The heat transfer coefficient increases as the mass flux increases because larger vapor velocities enhance convective heat transfer. In addition, a thinner liquid film is present between the vapor core and the wall at larger mass fluxes, thus increasing the thin film evaporation. Tube 2 gives the largest heat transfer coefficient when $G<400$ kg/(m^{2}s), while

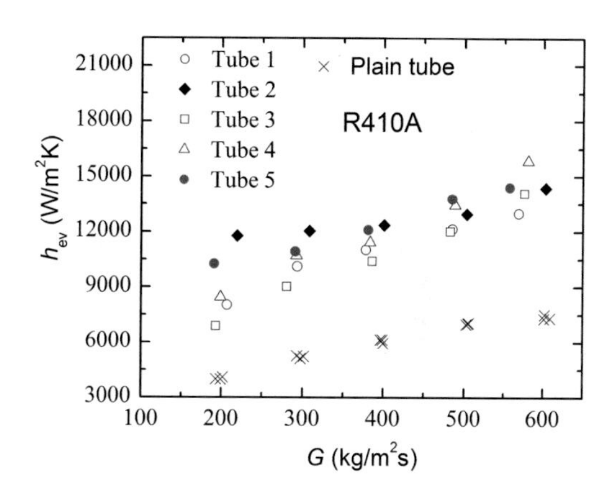

Fig. 9.13 Evaporation heat transfer coefficient versus mass velocity, with the inlet vapor quality of 0.1 and outlet vapor quality of 0.8. Wu et al. (2013b)

Table 9.1 Test tube geometries

Tube no.	d_o	d_i	n_s	α	β	e	A/A_{ni}	e/p_f
	mm	mm	–	deg	deg	mm	–	mm
Smooth tube	5.0	4.4	–	–	–	–	1.0	–
Tube 1	5.0	4.6	40	40	18	0.15	1.61	0.44
Tube 2	5.0	4.6	38	25	18	0.15	1.67	0.41
Tube 3	5.0	4.54	35	25	18	0.12	1.50	0.31
Tube 4	5.0	4.54	58	25	18	0.12	1.82	0.51
Tube 5	5.0	4.6	50	20	18	0.10	1.61	0.36
Tube 6	9.52	8.98	70	30	18	0.16	1.64	0.42

Tube 4 presents the best heat transfer performance at larger mass velocities. The ratio of the fin height e to the thickness δ of the liquid film between the vapor core and the wall affects the heat transfer performance. If e is far less than δ, the fins will be immersed in the liquid. Therefore, the liquid film becomes a thermal resistance and interfacial turbulence can be reduced. However, if e is far greater than δ, the increase in heat transfer area due to the increments of the fin height and number of starts is relatively inefficient in heat transfer enhancement. When the e/δ ratio is close to unity, fin tips will be covered by a very thin liquid film and periodic waves induced by the fins also make the liquid-vapor interface unstable, which maximizes thermal efficiency and interfacial turbulence, enhancing heat transfer greatly.

Microfin tubes are also efficient to enhance condensation heat transfer with a relatively low pressure drop penalty (Wu et al. 2014). Figure 9.14 presents the relationship between condensation heat transfer coefficient and mass flux G. For R410A condensation inside the five microfin tubes with the same outer diameter of 5 mm, Tube 4 has the best heat transfer performance covering almost all mass fluxes due to its largest area enhancement ratio A/A_{ni}, while Tube 3 performs worst when $G<600$ kg/(m^{2}s) due to its lowest area enhancement ratio A/A_{ni}. The heat transfer coefficient of the five microfin tubes: Tubes 1–5 is about 1.6–2.5 times of the smooth tube. Tube 6 has the lowest heat transfer coefficient because of its relatively larger inner diameter compared to the other five microfin tubes. The six microfin tubes have similar thermal performance when $G\geq 400$ kg/(m^{2}s), heat transfer coefficient increases with increasing mass flux. However, when $G<400$ kg/(m^{2}s), heat transfer coefficient decreases at first and then increases or flattens out gradually as mass flux decreases. In other words, mass flux has a non-monotonic relation with heat transfer coefficient in microfin tubes, with the possible reasons specified in Wu et al. (2014). The complex interactions between the microfins and the fluid could contribute to this mass-flux effect. The centrifugal force by the microfins spreads the liquid to the upper part, and surface tension pulls the condensate from the fin tips into the drainage channel at the base of the fins, forming a very thin liquid layer on the surface of microfins. Thus, heat transfer is enhanced greatly. Yang and Webb (1997) showed that the enhancement by liquid drainage decreases as the mass flux increases. In addition, the enhancement by interfacial turbulence is more significant

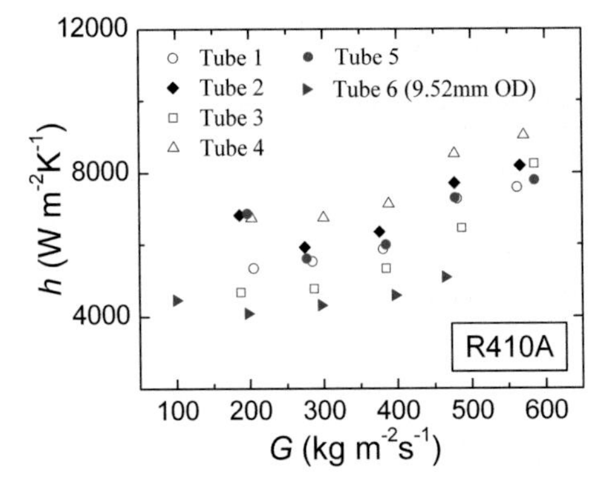

Fig. 9.14 Condensation heat transfer coefficient versus mass flux in microfin tubes with inlet and outlet vapor qualities of 0.8 and 0.1, respectively

at low mass fluxes and tends to be lower at larger mass fluxes. This is perhaps due to the enhanced turbulence already existing at large mass fluxes. Also Kedzierski and Goncalves (1999) presented that low Reynolds number flows may be enhanced more readily than high Reynolds number flows due to the reduction in the size of the turbulent eddies at the wall by the interaction of the flow with the fins because smaller eddies transfer momentum more efficiently than larger ones. In conclusion, the enhancement mechanism at small mass fluxes and suppression mechanism at relatively large mass fluxes may explain the complex mass-flux effect in microfin tubes. It is important to point out that the fin tip radius might affect the surface tension force greatly. A smaller fin tip radius provides a higher surface tension drainage force. More experimental investigations on a large variety of fin geometries, especially with focus on the fin tip radius, are needed to develop a robust model to predict the observed non-monotonic *h-G* phenomenon.

Herringbone tube (a typical microfin tube), if properly designed, can produce higher heat transfer coefficients than a helical microfin tube. A herringbone tube has a V shaped groove pattern that is expected to carry liquid in two different flow directions, creating liquid redistribution. The film is removed or supplied continuously by the V shaped groove. Figure 9.15 shows an example of a herringbone tube, which is produced using embossed, chevron-shaped microfins, with a portion of the fin-diverging at the bottom and another portion of the fin-converging at the top.

Besides microfins, other micro/miniscale elements, e.g., dimples, roughness elements and other special enhancement patterns of different shapes can also be incorporated to augment condensation and evaporation heat transfer. Surface patterns of the enhanced heat transfer (EHT) tube produced by Vipertex is shown in Fig. 9.16. Enhancement characters are made up of dimples and petals, as shown in Fig. 9.16. The EHT tube is neither a classic "integral roughness" tube with little surface area increase nor an internally finned tube with a surface area increase and no flow

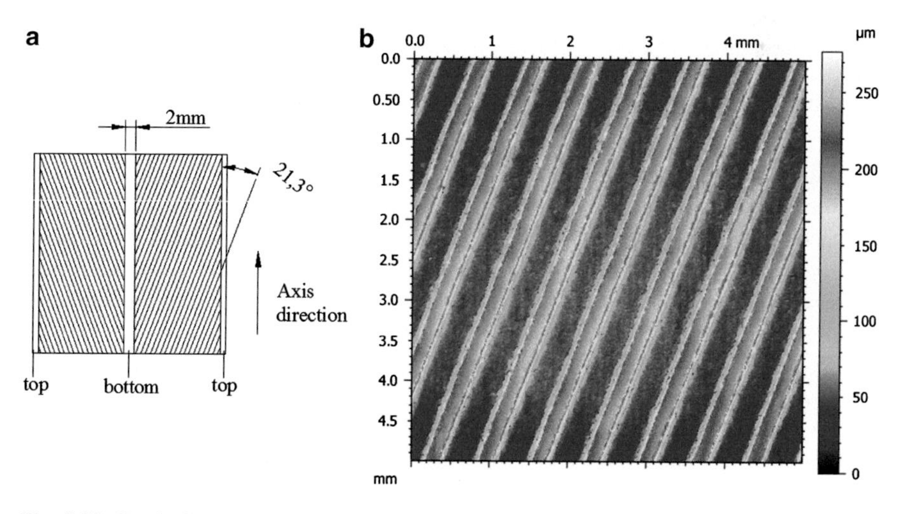

Fig. 9.15 Herringbone tube configuration: (**a**) herringbone pattern, (**b**) false color view for the herringbone profile. Guo et al. (2015)

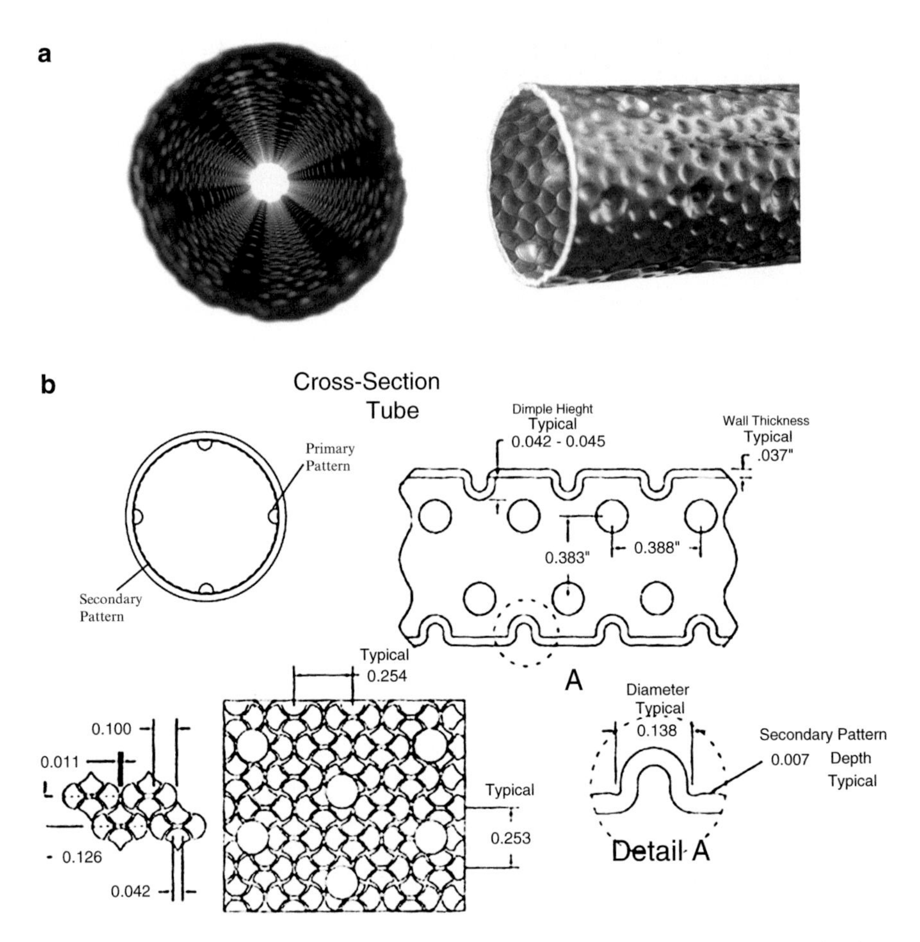

Fig. 9.16 EHT tube (**a**) inner surface and outer surface enhancement structure, (**b**) details of the primary enhancement structure with a typical diameter of 3.505 mm (0.138 in.) and a dimple height of 1.067–1.143 mm (0.042–0.045 in.), and secondary petal-shape patterns with a typical diameter of 2.54 mm (0.1 in.) and a typical height of 0.178 mm (0.007 in.). Guo et al. (2015)

separation. The EHT surface is more of a hybrid surface with a surface area increase and flow separation produced by the primary dimple/protrusion enhancement. This enhanced heat transfer surface produces a combination of more nucleation sites, increased turbulence, boundary layer disruption, secondary flow generation and increased heat transfer surface area (area determined using the optical profiler), all leading to heat transfer enhancement for a wide range of conditions.

Figure 9.17 presents the relationship between the condensation heat transfer coefficient and mass flux for smooth, herringbone and EHT tubes using refrigerants: (a) R22; (b) R32 and (c) R410A. The three tubes has nearly the same inner diameter of 11.43 mm and the same outer diameter of 12.7 mm. For all the three refrigerants, the herringbone tube presents the highest heat transfer coefficient while the smooth tube produces the lowest heat transfer coefficient for all tested mass fluxes. The heat transfer coefficient of the herringbone tube is around 2.0–3.0 times that of the smooth tube for

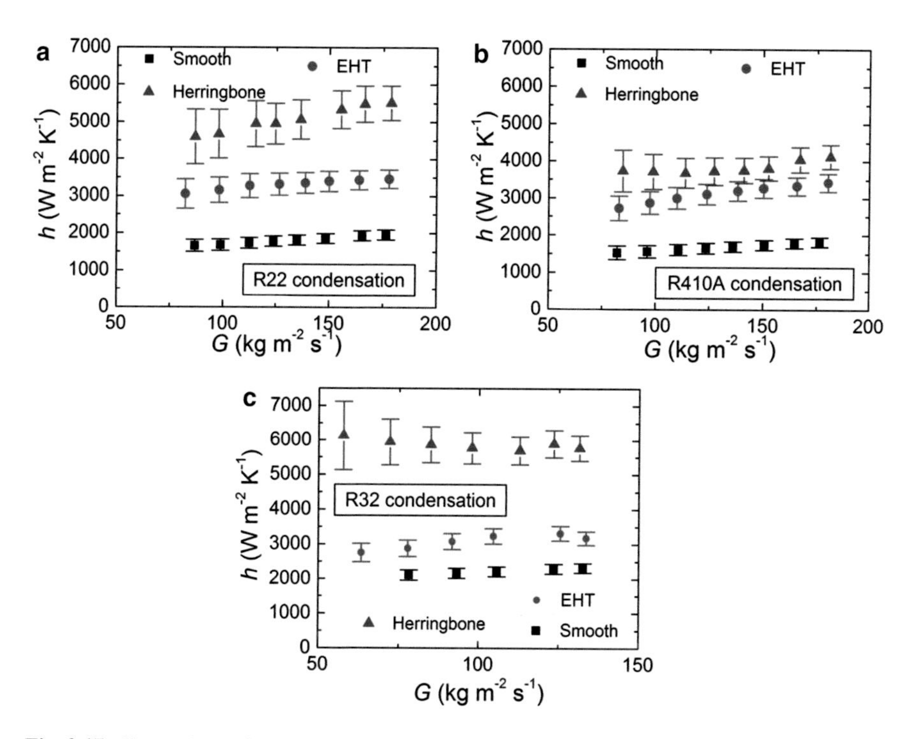

Fig. 9.17 Comparison of the condensation results for smooth, herringbone and enhanced surface EHT tubes for the following conditions: saturated temperature, inlet vapor quality and outlet vapor quality are (**a**) 47.0±0.5 °C, 0.80±0.01, 0.20±0.01, (**b**) 47.0±0.5 °C, 0.80±0.01, 0.20±0.01, and (**c**) 45.0±0.5 °C, 0.80±0.01, 0.20±0.01, respectively. Guo et al. (2015)

condensation. Since the heat transfer enhancement ratio (2.0–3.0) is larger than the inner surface heat-transfer area ratio (1.57), heat transfer enhancement from mechanisms other than surface area increase must come into play. For condensation, there are mainly two enhancement mechanisms besides the surface area increase. In the tested herringbone tube, the diverging fins at the bottom push the liquid up to the top (i.e., redistribute the liquid film more evenly), thus enhancing the heat transfer. In addition, the liquid converging at the top induces liquid mixing and turbulence, and this might also play an important role in enhancing heat transfer. The EHT tube also produces enhanced condensation heat transfer performance, with the heat transfer coefficient in the range of 1.3–1.95 times that of the smooth tube. The heat transfer enhancement ratio (1.3–1.95) is larger than the inner surface heat-transfer area ratio (1.112). In addition to surface area increase, enhanced heat transfer is produced by the increased turbulence and flow separation produced by the primary dimples; additionally boundary layer disruption and flow mixing is produced by the secondary petal pattern.

Figure 9.18 compares the experimental data of the in-tube evaporation heat transfer coefficient and mass flux for smooth, herringbone and EHT tubes using refrigerants (a) R22; (b) R32 and (c) R410A. For all the three refrigerants, the EHT tube presents the highest heat transfer coefficient while the smooth tube produces the lowest heat transfer coefficient for all mass fluxes. Despite a large increase in the

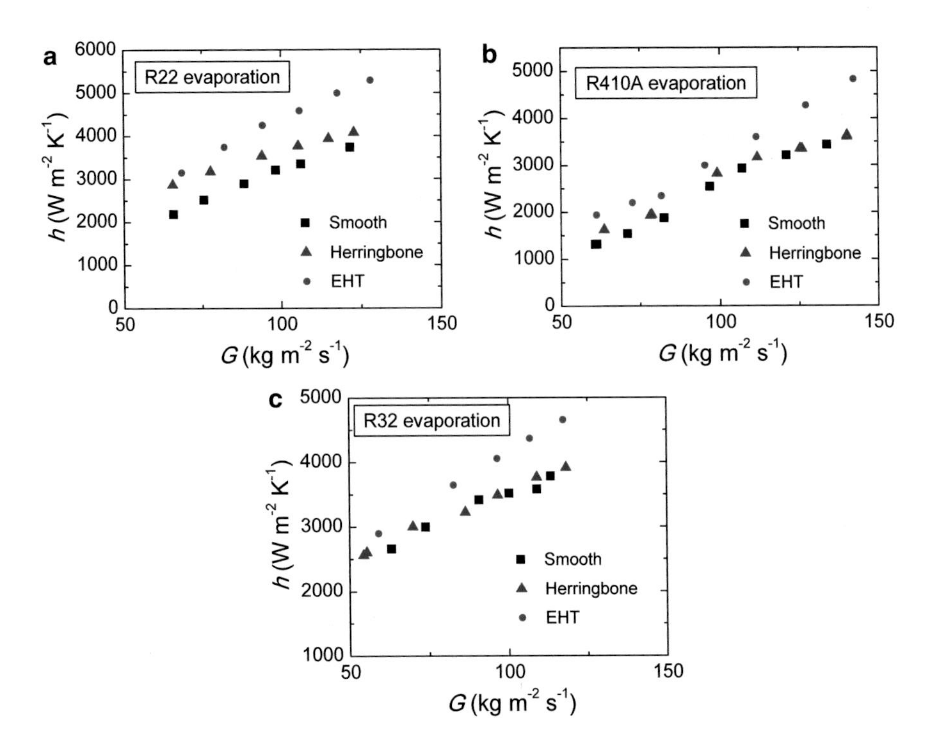

Fig. 9.18 Comparison of the evaporation results for smooth, herringbone and enhanced surface EHT tubes for the following conditions: saturated temperature, inlet vapor quality and outlet vapor quality are (**a**) 6.0±0.2 °C, 0.10±0.02, 0.90±0.02, (**b**) 6.0±0.2 °C, 0.10±0.02, 0.90±0.02, and (**c**) 6.0±0.2 °C, 0.10±0.02, 0.80±0.02, respectively. Guo et al. (2015)

inner surface area, the herringbone tube does not present an obvious heat transfer enhancement. As can be seen in Fig. 9.18, the evaporation heat transfer coefficient of the herringbone tube is only slightly larger than that of the smooth tube. Heat transfer enhancement ratios in the herringbone tube are far less than the surface area ratio (1.57) for all mass fluxes (except for R22 at low mass fluxes), which means that not all the inner surface area participates in heat transfer. Part of the fin surface is exposed to the vapor, and therefore is inefficient for heat transfer. Liquid film dryout might occur for evaporation (this is especially true at high vapor qualities), since the saturation temperature is much lower than that for condensation and the vapor density is much lower. In addition, the large latent heat of vaporization and the large vapor density tend to form thin liquid films. The liquid film thickness is less than half of the fin height when the vapor quality is larger than 0.5, which is calculated using the film height equation developed by Wellsandt and Vamling (2005). All these factors result in a low evaporative heat transfer enhancement for the herringbone tube. Among the three tubes, the EHT tube gives the best heat transfer performance for all the three refrigerants. The heat transfer enhancement ratio (1.2–1.4) is larger than the inner surface area ratio (1.112). The inner surface area increase of the EHT tube enhances heat transfer; in addition the dimple and petal arrays in the enhanced surface structure further enhances the heat transfer performance of the EHT tube due to more nucleation sites, increased interfacial turbulence, boundary layer disruption, flow separation and secondary flow generation.

Corrugated plates are routinely used in plate heat exchangers (PHE) to enhance heat transfer. An example of a plate heat exchanger and a corrugated plate is shown in Fig. 9.19. The plate surface corrugations readily promote enhanced heat transfer by means of several mechanisms that include promoting swirl or vortex flows, disruption

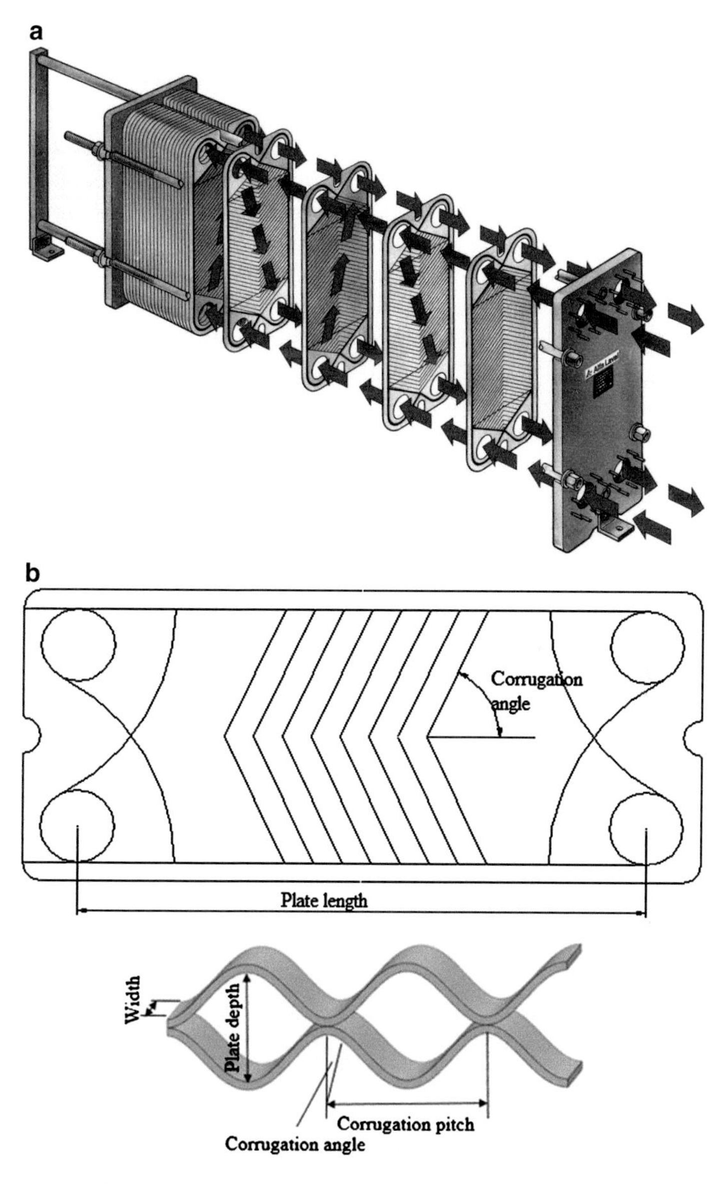

Fig. 9.19 (**a**) Plate heat exchanger, and (**b**) schematic of a corrugated plate

and reattachment of boundary layers, small hydraulic diameter flow passages, and increased effective heat transfer area (Wang et al. 2007a, b). Flow boiling and condensation heat transfer characteristics in plate heat exchangers have been presented in the literature (e.g., Wang et al. 1999; Hsieh and Lin 2002; Kuo et al. 2005; Rao Bobbili et al. 2006; Sterner and Sundén 2006; Djordjevic and Kabelac 2008: Lee et al. 2013; Longo et al. 2014). Figure 9.20 shows the quasi-local heat transfer coefficient as a function of the heat flux at a saturation temperature of 283 K for the lower mass flux of 10 kg/(m²s) and a saturation temperature of 273 K for the higher mass flux of 18 kg/(m²s). This diagram shows that the nucleate boiling and the convective boiling regimes can be identified to some extend within PHEs.

Micro-pin-fins of various cross-sections can be fabricated on the surface to enhance heat transfer. Short fin lengths are preferred to disrupt the thermal boundary layer and obtain repeated developing flow. The small fin height can maintain high fin efficiency. Flow mixing and flow uniformity are greatly improved by using interrupted microfins. However, consideration should be taken for the increased pressure drop. The fin height and shape need to be controlled to maintain a relatively low pressure drop penalty. Ma et al. (2009) studied flow boiling heat transfer of FC-72 over micro-pin-finned surfaces. Two micro-pin-finned chips with the same fin thickness of 30 μm and two different fin heights of 60 μm (PF30-60) and 120 μm (PF30-120), respectively, were tested. SEM images of the two micro-pin-finned chips are shown in Fig. 9.21. Figure 9.22 shows the comparison between the boiling curves for all surfaces at $\Delta T_{sub}=25$ K. The slopes of boiling curves and the critical heat fluxes increase in the order of chip S (smooth chip for comparison), PF30-60 and PF30-120 at the same velocity, indicating that all the

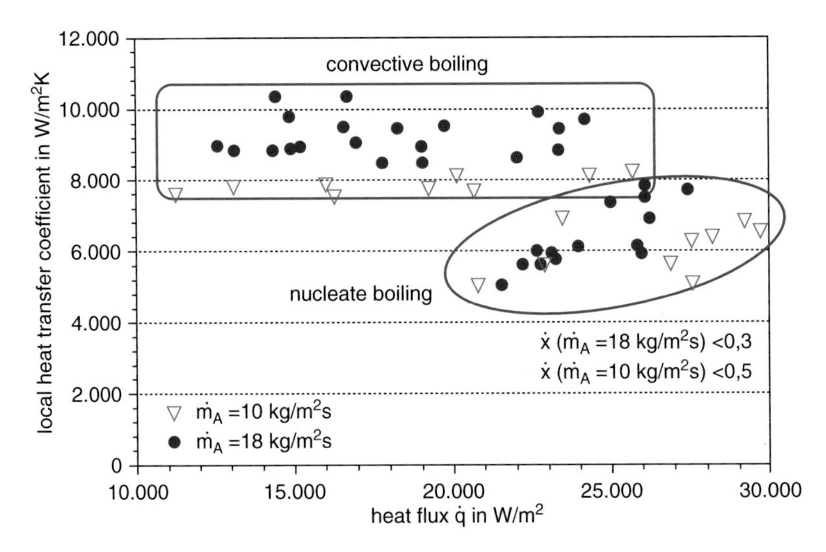

Fig. 9.20 Quasi-local heat transfer coefficients for ammonia shown as a function of heat flux. Two different mass fluxes are shown. Both convective and nucleate boiling mechanisms can be identified. Djordjevic and Kabelac (2008)

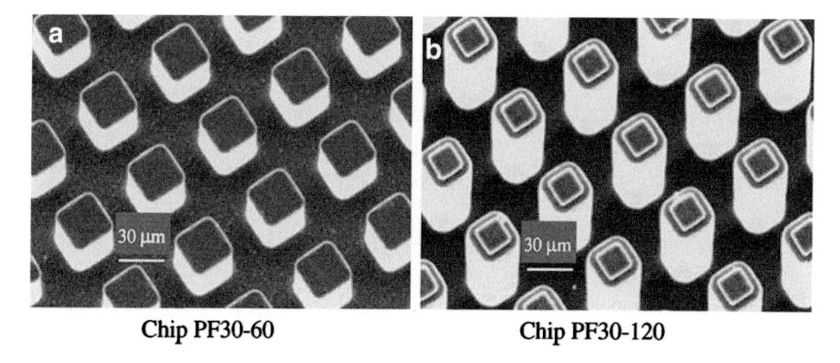

Fig. 9.21 SEM images of micro-pin-fins. Ma et al. (2009)

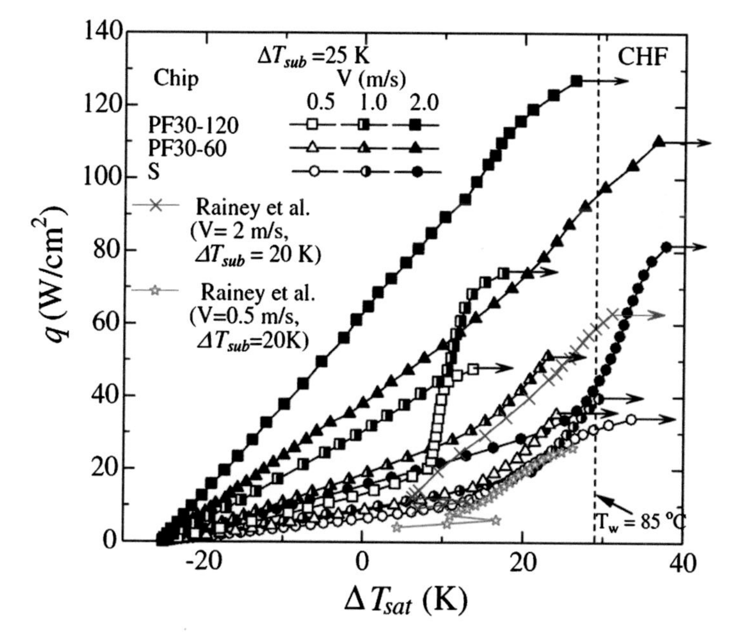

Fig. 9.22 Comparison of boiling curves of FC-72 for micro-pin-finned chips and a smooth chip at a subcooling of 25 K. Ma et al. (2009)

micro-pin-finned surfaces have considerable heat transfer enhancement compared to the smooth one. The results of Rainey et al. (2001) are also shown for comparison. The growth and movement of the bubbles in the confined gaps between fins can bring about micro-convection and form thin liquid layer for evaporation, which makes the fin profiles become effective heat transfer area, creating heat transfer enhancement. Observation on boiling phenomena on the micro-pin-finned chip reveals that this surface can create more active nucleation sites and make the bubbles rest on the surface for a longer time to evaporate, thus increasing the heat transfer performance.

9.3 Insert Devices

Insert devices involve various geometric forms that are inserted into a flow channel. There are various types of insert devices, such as wire coil insert, twisted-tape insert (vortex generator), helical vane insert, extended surface insert, perforated insert and star insert.

Celata et al. (1994) found that helically coiled wires can be used as turbulence promoters to enhance CHF with respect to a smooth channel. Helically coiled wires (d = 1.0 mm, pitch = 20.0 mm) allowed an increase in CHF up to 50 %, coupled with a moderate increase in pressure drop (less than 20 %). No observable influence of the channel orientation was detected. Ma et al. (2004) investigated the film condensation in a vertical, internally finned tube with a wire insert. As shown in Fig. 9.23, the wire insert considerably improves the heat transfer of the plain tube. However, the addition of the wire insert in the finned tube only improves the heat transfer performance slightly. With regard to a relatively large pressure drop penalty due to the wire insert, it is thus not recommended to use wire inserts in enhanced tubes for condensation.

Hsieh et al. (2003) experimentally investigated heat transfer of R134a and R600a in horizontal tubes with vertically positioned perforated strip-type inserts. The test section is a 2000 mm long, smooth copper tube with an inner diameter of 10.6 mm, inserted with perforated strip-type copper inserts. The flow patterns were identified

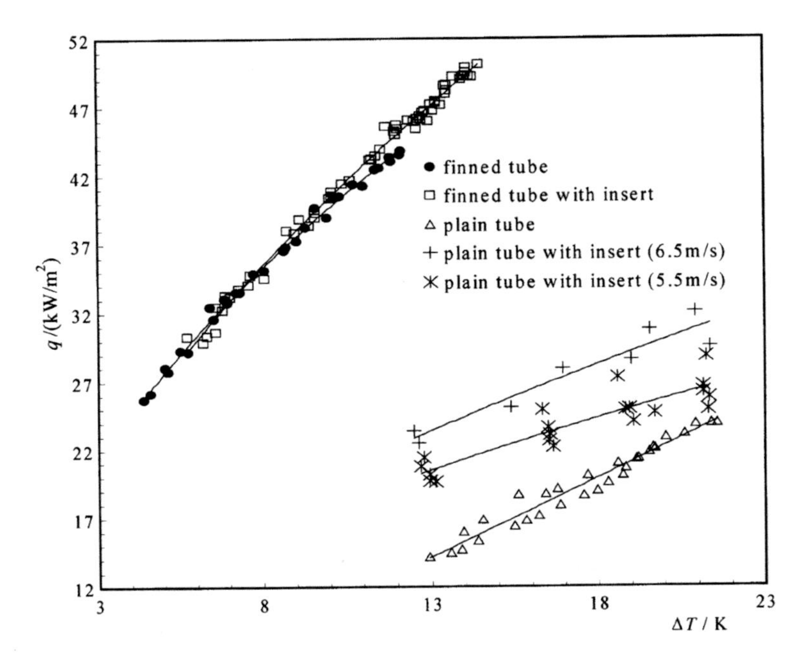

Fig. 9.23 Heat flux vs. vapor-side temperature difference for plain and finned tubes with and without wire inserts. Ma et al. (2004)

for different test tubes and flow conditions. The heat transfer performance and pressure drop can be improved up to 2.5 and 1.5, respectively, for the tube with a 96-hole perforated strip insert. All comparisons were based on the same nominal mass flow rate. As shown in Fig. 9.24, the perforated insert introduces an earlier transition from intermittent flow to annular flow. The heat transfer enhancement might be mainly caused by the enhancement in convective evaporation.

Twisted-tape insert generates helical swirling motion inside a tube to significantly enhance convective heat transfer. The primary mechanism entails imparting a centrifugal force component to the longitudinal fluid motion, which superimposes secondary circulation over the main axial flow to promote cross-stream mixing

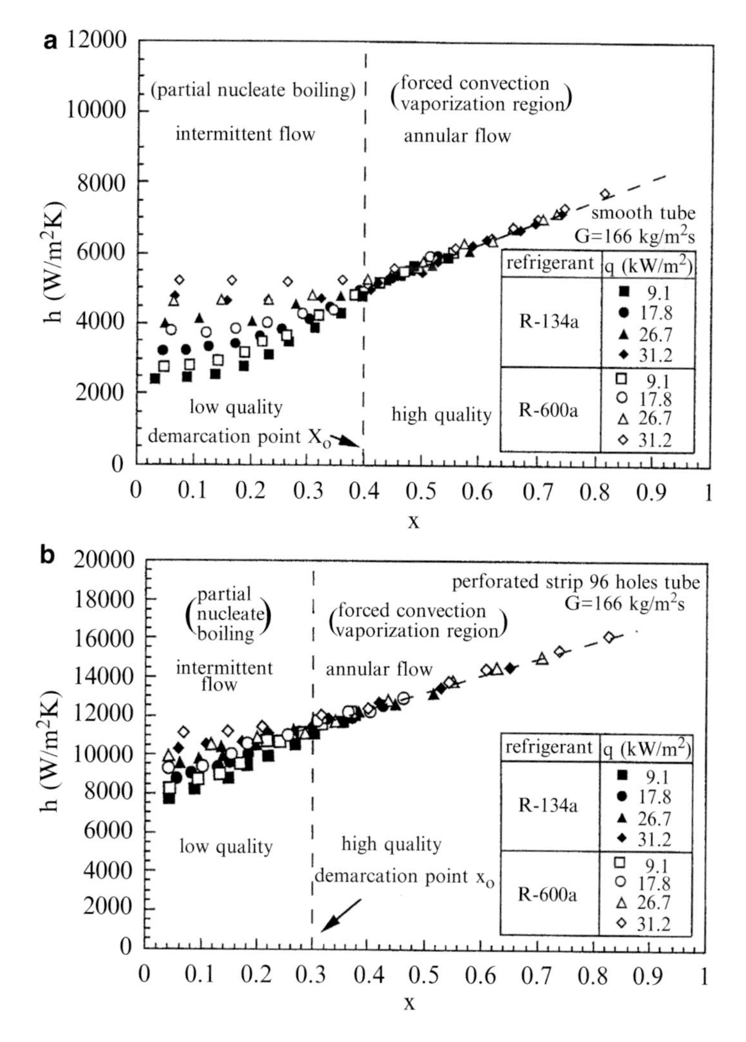

Fig. 9.24 Local heat transfer coefficient vs. quality at constant mass flux and different heat fluxes in (**a**) a smooth tube and (**b**) a tube with a 96-hole perforated strip insert. Hsieh et al. (2003)

(Hata and Masuzaki 2011; Manglik and Bergles 2013). In flow boiling with net vapor generation, tape-twist-induced helical swirl pushes liquid droplets from the tube core to the wall to enhance heat transfer and delay dryout. In subcooled boiling, the radial pressure gradient due to the swirl promotes vapor removal from the heated surface to retard vapor blanketing and accommodate higher heat fluxes (Lopina and Bergles 1973).

9.4 Curved Geometry

Curved geometry might enhance heat transfer of two-phase flow as an additional centrifugal force will be generated in the curved geometry (Owhadi et al. 1968; Garimella et al. 1988; Zhao et al. 2003; Mozafari et al. 2015). Examples of curved geometries are wavy channels, helically coiled tubes, serpentine tubes and bends etc. Santini et al. (2016) experimentally studied flow boiling of water in a 24 m long full-scale helically coiled steam generator tube, prototypical of the steam generators with in-tube boiling used in small modular nuclear reactor systems. The heat transfer coefficient was found to depend on the mass flux and the heat flux, indicating that both nucleate boiling and convective boiling are contributing to the heat transfer process. Results of the helically coiled tubes can be well fitted by seven widely quoted flow boiling correlations for straight tubes, with a mean absolute percentage error within 15–20 %, which was comparable with the experimental uncertainty of the measured heat transfer coefficient values. Therefore, it seems that curvature effects on flow boiling are small and negligible in practical applications.

9.5 Additives

Additive is another method to enhance heat transfer of two-phase flow. Nanoparticles, micro/nano-encapsulated phase-change materials, surfactants are common examples of additives (Kim et al. 2001; Yang and Liu 2012; Kumar et al. 2016). There are several possible reasons for the heat transfer enhancement by additives. Firstly, additive can change the thermophysical properties of the heat transfer fluid. For example, nanofluids, engineered colloidal suspensions of nanoparticles of a base fluid, generally provide higher thermal conductivity compared to the base fluid due to the relatively large thermal conductivity of the added nanoparticles. Surfactants change the viscosity and surface tension of heat transfer fluids and thus affect the fluid flow and bubble/droplet dynamics. Secondly, additives might modify the wettability of the heat transfer surface and/or the surface morphology. For instance, nanoparticles may deposit on the surface and therefore modify the surface morphology. In addition, additives may present additional heat transfer mechanisms. For example, at some cases the Brownian motion of nanoparticles may be comparably significant in enhancing heat transfer. The presence of surfactants at the liquid-vapor interface in two-phase flow might cause agitation at the surface (interfacial turbulence).

Peng et al. (2011) experimentally investigated the boiling heat transfer performance of refrigerant-based Cu/R113 nanofluids. Refrigerant R113 is chosen as the base fluid because R113 is in liquid state at room temperature and atmospheric pressure. Therefore it is easy to disperse the nanoparticles into R113 evenly. Most of widely used refrigerants, e.g., R410A and R134a, are in vapor state at room temperature and atmospheric pressure, and it is difficult to prepare nanofluids with well-dispersed nanoparticles based on such refrigerants. It can be seen from Fig. 9.25 that the heat transfer oefficient of Cu/R113 nanofluid is larger than that of pure R113. The maximum enhancement of the heat transfer coefficient (up to 55.4 %) occurs at the highest nanoparticle concentration. The heat transfer enhancement of the refrigerant-based nanofluid is said to be mainly caused by (a) the change of the heating surface characteristics due to the interaction between nanoparticles and the heating surface, and (b) the change of the thermophysical properties.

Kim et al. (2010a, b) compared the subcooled flow boiling heat transfer performance of dilute alumina, zinc oxide, and diamond nanofluids at atmospheric pressure. Figure 9.26 summarizes the heat transfer coefficient results at $G = 1500$ and 2500 kg/m^{2}s, respectively. The heat transfer coefficient ratios of nanofluid to water are predominantly located within ±20 % around unity, which is the typical magnitude of the uncertainties in boiling heat transfer analysis. It was found that the values of the nanofluid and water heat transfer coefficient are similar (within ±20 %) for comparable test conditions. A confocal microscopy-based examination of the test section revealed that nanoparticle deposition on the boiling surface occurred during nanofluid boiling (Fig. 9.27). Such deposition changes the number

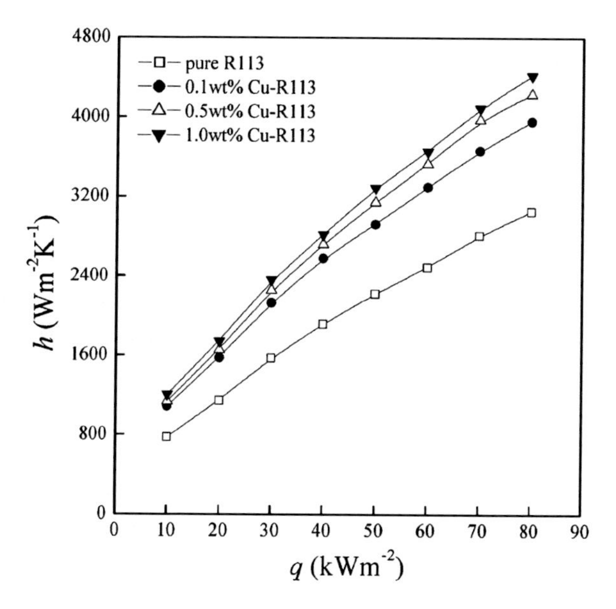

Fig. 9.25 Heat transfer coefficient of Cu/R113 nanofluid versus heat flux. Peng et al. (2011)

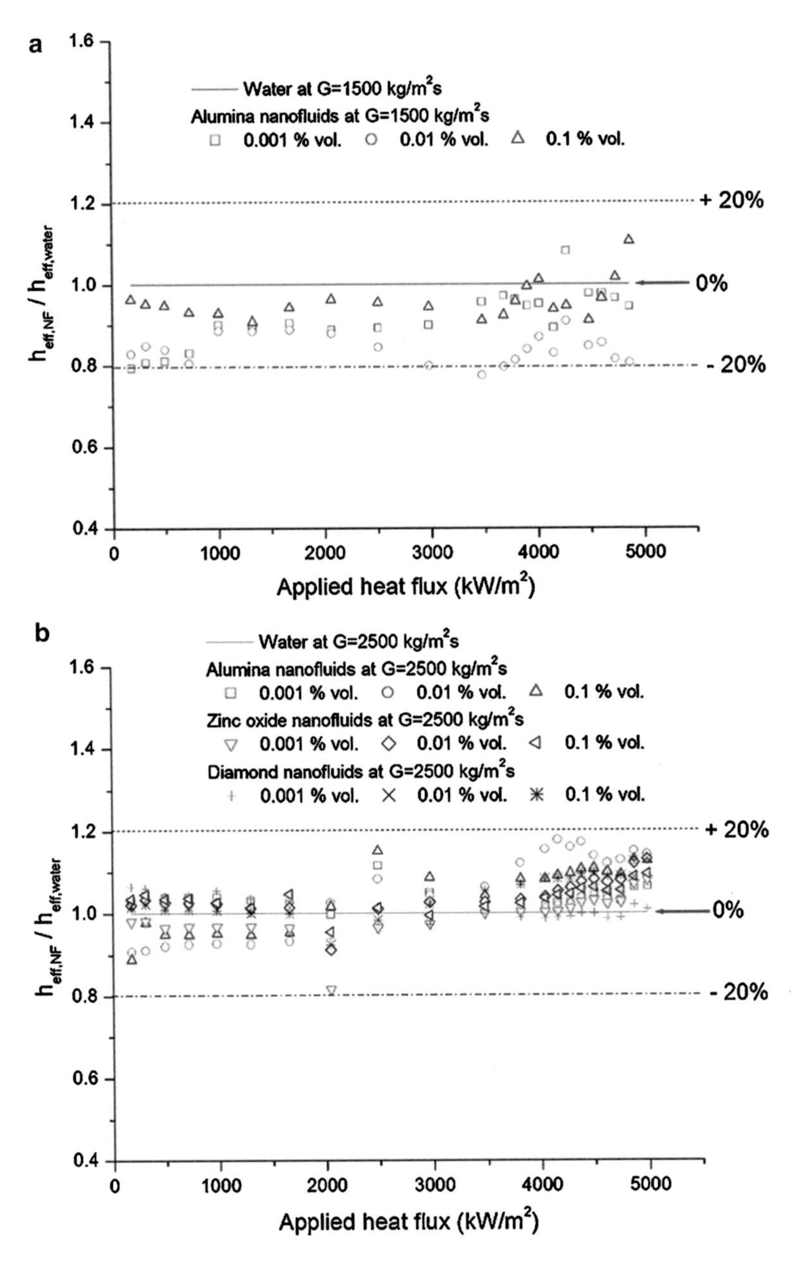

Fig. 9.26 Comparison of heat transfer coefficient ratios of nanofluid to water at mass flux of (**a**) 1500 kg/m^{2}s, and (**b**) 2500 kg/m^{2}s. Kim et al. (2010a, b)

of micro-cavities on the surface, but also changes the surface wettability. The increase in nucleation site density due to nanoparticle deposition enhances heat transfer. However, the enhancement might be offset by other possible effects, such as changes in bubble departure diameter and departure frequency.

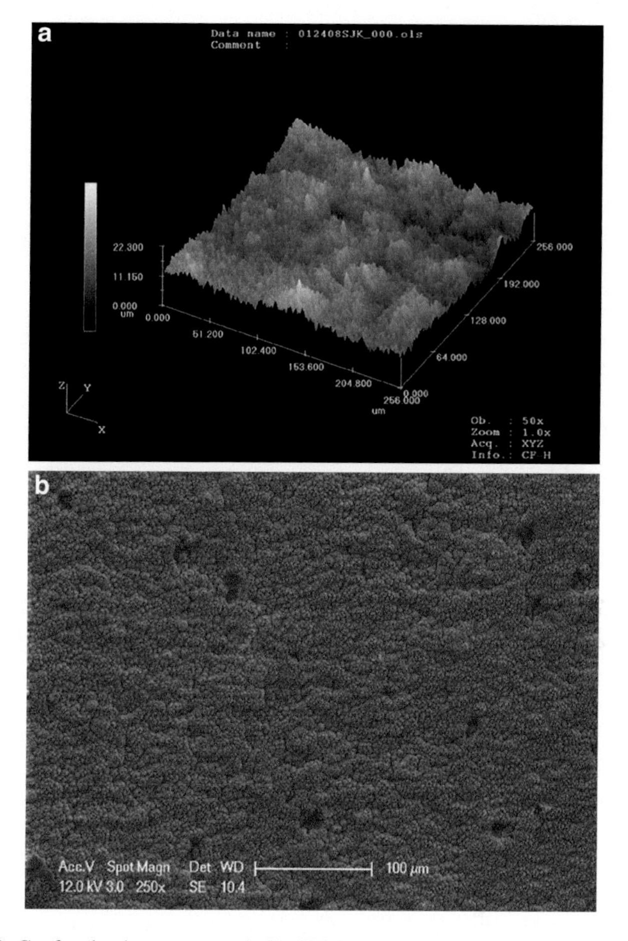

Fig. 9.27 (**a**) Confocal microscopy and (**b**) SEM images of the test section surface for the 0.1 vol.% alumina nanofluid run. Kim et al. (2010a, b)

In contrast to Kim et al. (2010a, b), Yu et al. (2015) tested flow boiling heat transfer in a minichannel and observed that the addition of alumina nanoparticles into water delays the onset of nucleate boiling (ONB), and the extent of delay is proportional to the nanoparticle concentration. Several mechanisms responsible for the ONB delay in flow boiling of nanofluids were given in Yu et al. (2015) and the mechanisms are listed as follows:

1. The nanoparticle layer deposited on the wall surface may alter the profile of active nucleation sites. It was reported that once boiling occurs, nanoparticles first aggregate at the liquid–vapor interface, and then quickly adhere to the wall surface (Lee and Mudawar 2007). Figure 9.28 shows a possible distribution of nucleation sites on a boiling surface after nanoparticle deposition. Some microscale cavities might be filled by the nanoparticle aggregates. Consequently, the size range and number density of available nucleation sites will diminish, making the ONB more difficult to occur.

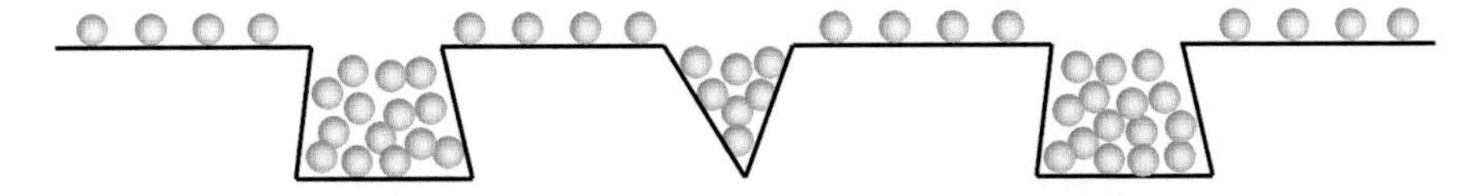

Fig. 9.28 An example of nanoparticle deposition on the boiling surface

2. The development of thermal boundary layer is retarded in nanofluids, i.e., the thermal boundary layer becomes thinner in nanofluids, due to the effect of the shear-induced nanoparticle migration on the effective viscosity and thermal conductivity (Nnanna 2007; Liu and Yu 2011). In the classical ONB model, Hsu (1962) showed that the size range of theoretically eligible nucleate cavities depends proportionally on the thickness of the thermal boundary layer. Therefore, suppressed thermal boundary layer will result in less active nucleation sites in flow boiling of nanofluids.
3. ONB is strongly dependent on the wettability of the boiling surface. As nanoparticles deposit on the surface, the contact angle decreases and thus the size range of active nucleation sites diminishes.

Different from nanofluid works such as Lee and Mudawar (2007), Kim et al. (2010a, b), Vafaei and Wen (2010) and Yu et al. (2015), Xu and Xu (2012) did not observe nanoparticle deposition phenomenon for flow boiling of alumina/water nanofluid (of a low weight concentration of 0.2 %) in a single microchannel. It is found that nanofluid significantly mitigate the flow instability without nanoparticle deposition effect. As shown in Fig. 9.29, large oscillations occurred in flow boiling of water but not in flow boiling of nanofluid. Nanofluid delays the onset of flow instability and thus stabilizes the boiling flow. In microchannels, thin film evaporation is one of the major heat transfer mechanisms in bubbly flow, elongated flow and annular flow (Thome et al. 2004; Wang et al. 22007a, b; Wu et al. 22013a, b; Wu and Sundén 2015b, c). Xu and Xu (2012) proposed that nanoparticle additives enhance thin film evaporation. Figure 9.30 shows the three-phase contact line for pure water and nanofluid. There is one more region compared with water. Region IV is the thin liquid film evaporation heat transfer region with nanoparticles involved. Comparing Fig. 9.30a and b, the nanofluid decreases region I and increases the percentage of the thin liquid film region with nanoparticles involved on the heater surface, improving the heat transfer performance for nanofluid. Besides, as indicated in Xu and Xu (2012), there are two other factors account for the heat transfer enhancement by nanofluid: (a) the bubble departure size is smaller for nanofluid than that for water, inducing quick miniature bubble departure and disturbing the boundary layer frequently. (b) The dry area for region I is shortened by the nanofluid, improving the heat transfer.

In general, flow boiling of nanofluid is a rather complex phenomenon. There is still disagreement on whether nanofluid can enhance flow boiling heat transfer, not to mention the underlying heat transfer mechanisms when nanoparticles are involved. At least, most studies agreed improvement in CHF (e.g., Li and Peterson 2007; Kim et al. 2008, 2010a, b; Furberg et al. 2009; Henderson et al. 2010; Wu and

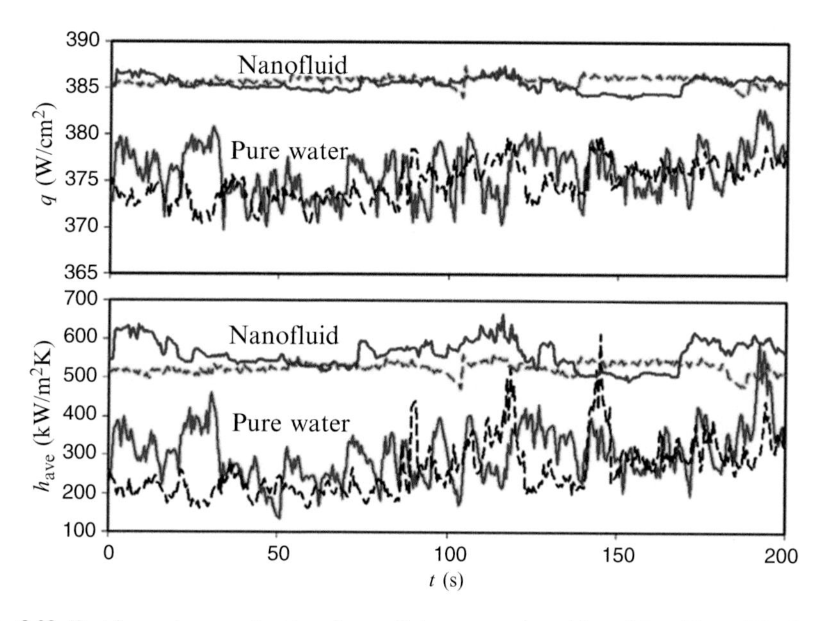

Fig. 9.29 Heat flux and average heat transfer coefficient versus time. Adapted from Xu and Xu (2012)

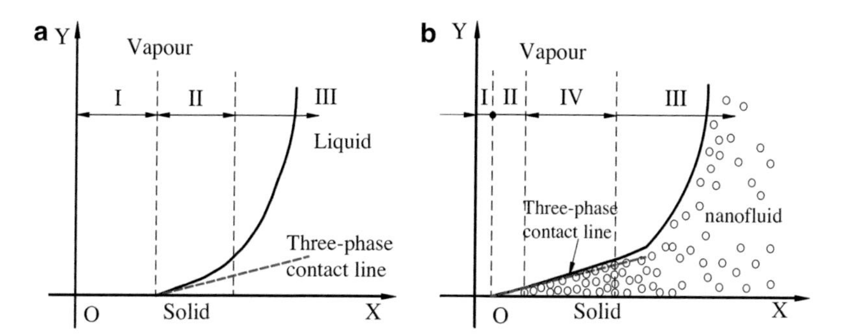

Fig. 9.30 Solid-liquid-vapor three phase contact line for (**a**) pure water, and (**b**) nanofluid. Xu and Xu (2012)

Sundén 2014) and have established that the CHF enhancement by nanofluids results from a thin porous nanoparticle deposition layer on the heater surface which serves to improve the wettability and capillarity of the boiling surface. The main disadvantages of using nanofluids for boiling heat transfer improvement are the nanoparticle pollution and poor nanoparticle deposition strength (Krishna et al. 2011; Bi et al. 2015). As the progress in nanoengineering, nanofluid boiling has inspired nanoscale surface coatings to enhance heat transfer in two-phase flow, without a large change in the surface topography at microscale. Please see above for heat transfer enhancement by nanoscale coatings.

Another common additive to enhance phase-change heat transfer is surfactants. Surfactants are compounds that lower the surface tension between two liquids or between a liquid and a solid. Dropwise condensation might be established by adding proper chemicals to the vapor or treating the surface with proper chemicals, such as oils (Sundén 2012). The upward flow boiling heat transfer of water and ethylene glycol/water (EG/W) mixtures in a vertical mini-tube (2.3 mm) was experimentally determined by Feng et al. (2016). Sodium dodecylbenzenesulfonate (SDBS) was used as a surfactant. The results of EG/W and 300 ppm SDBS solutions at mass fluxes of 1070 and 640 kg/(m²s) are presented in Fig. 9.31. For 1070 kg/(m²s), after a sharp increase at low effective qualities, the local two-phase heat transfer coefficient of EG/W and 300 ppm SDBS solutions both drops and converges to two different groups. For EG/W, the heat transfer deterioration along the tube occurs when the nucleation heat transfer is mainly suppressed by lower

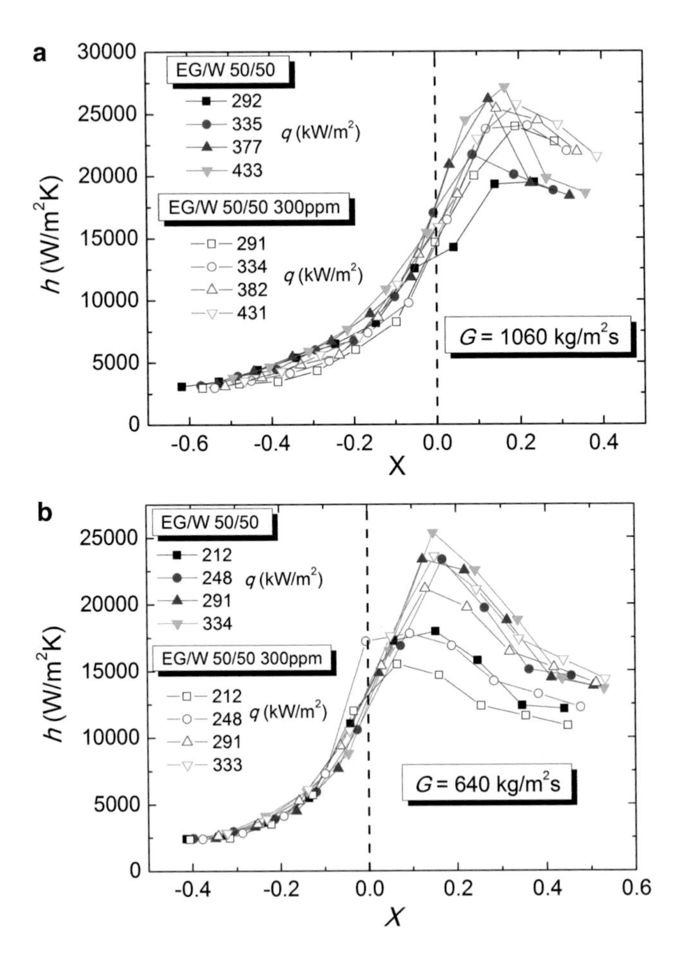

Fig. 9.31 Heat transfer coefficients of EG/W mixture and 300 ppm SDBS solutions for different heat fluxes (kW/m²): (**a**) higher mass flux, (**b**) lower mass flux. Feng et al. (2016)

local reduced pressure and exhausted nucleation sites. While the EG/W data exhibits a sharp decrease of HTC after the apex, the SDBS solutions maintains a relatively high HTC. The addition of trace SDBS remarkably enhances the nucleate boiling heat transfer after the critical effective quality. The surfactant SDBS promotes nucleation and evaporation, and therefore enhances the heat transfer. For 640 kg/(m^2s), the trend is somewhat different. Generally, the boiling heat transfer coefficient of 300 ppm SDBS solutions is lower than that of EG/W. The deterioration in HTC results from intermittent dryout. Surfactant additives promote vigorous nucleation and therefore may also lead to early intermittent dryout.

Chapter 10
Active Techniques

Active techniques require external power, such as electric or acoustic fields, surface/fluid vibrations, suction and jet impingement. In general, active techniques are not as common as passive techniques in industry as active techniques need external energy. The majority of commercially interesting techniques are mainly limited to passive techniques. However, active techniques such as electrohydrodynamic (EHD) enhancement of boiling and condensation indicate significant potential (Akira and Hiroshi 1988; Seyed-Yagoobi and Bryan 1999). EHD enhancement of heat transfer in two-phase flow features several advantages: (a) able to control the heat transfer coefficient by changing the applied voltage, (b) significant enhancement, (c) no moving parts, and (d) the electric power input is usually negligible. The heat transfer enhancement is highly dependent on quality, flow regime, heat flux, mass flux, and the strength of the radial EHD forces relative to the flow axial momentum. The EHD phenomena involve the interaction of electric fields and flow fields in a dielectric fluid medium. This interaction, under certain conditions, results in electrically induced fluid motion and/or interfacial instabilities, which are caused by an electric body force. When this force is enhancing heat transfer it is thinning and/or destabilizing the liquid layer, depending on the mass flux. However, this force can thin the liquid layer to a point of removing it and can drastically reduce the heat transfer, especially at low mass fluxes and high heat fluxes (Bryan and Seyed-Yagoobi 2001). The electric body force density acting on the molecules of a fluid in the presence of an electric field consists of three terms, as shown below (Melcher 1981):

$$\mathbf{f}_e = \rho_e \mathbf{E} - \frac{1}{2} E^2 \nabla \varepsilon + \frac{1}{2} \nabla \left[E^2 \rho \left(\frac{2\varepsilon}{\partial \rho} \right)_T \right] \tag{10.1}$$

The three terms in Eq. (10.1) stand for the electrophoretic, dielectrophoretic, and electrostrictive components of the electric force. For two-phase flows, the dielectrophoretic force dominates because the gradient in the dielectric permittivity, ∇ε, is very high at the vapor-liquid interface, resulting in a large EHD force acting on the

S.K. Saha et al., *Advances in Heat Transfer Enhancement*,
DOI 10.1007/978-3-319-29480-3_10

interface. This force can cause interfacial instabilities that force the liquid with higher permittivity to move to the regions of higher electric field. This phenomenon is usually referred to as the liquid extraction phenomenon and is believed to be the primary mechanism responsible for flow regime transitions which cause heat transfer enhancement. Sadek et al. (2006) studied in-tube condensation of R134a in a horizontal, single-pass, counter-current heat exchanger with a rod electrode placed in the centre of the tube. As shown in Fig. 10.1, the heat transfer coefficient was

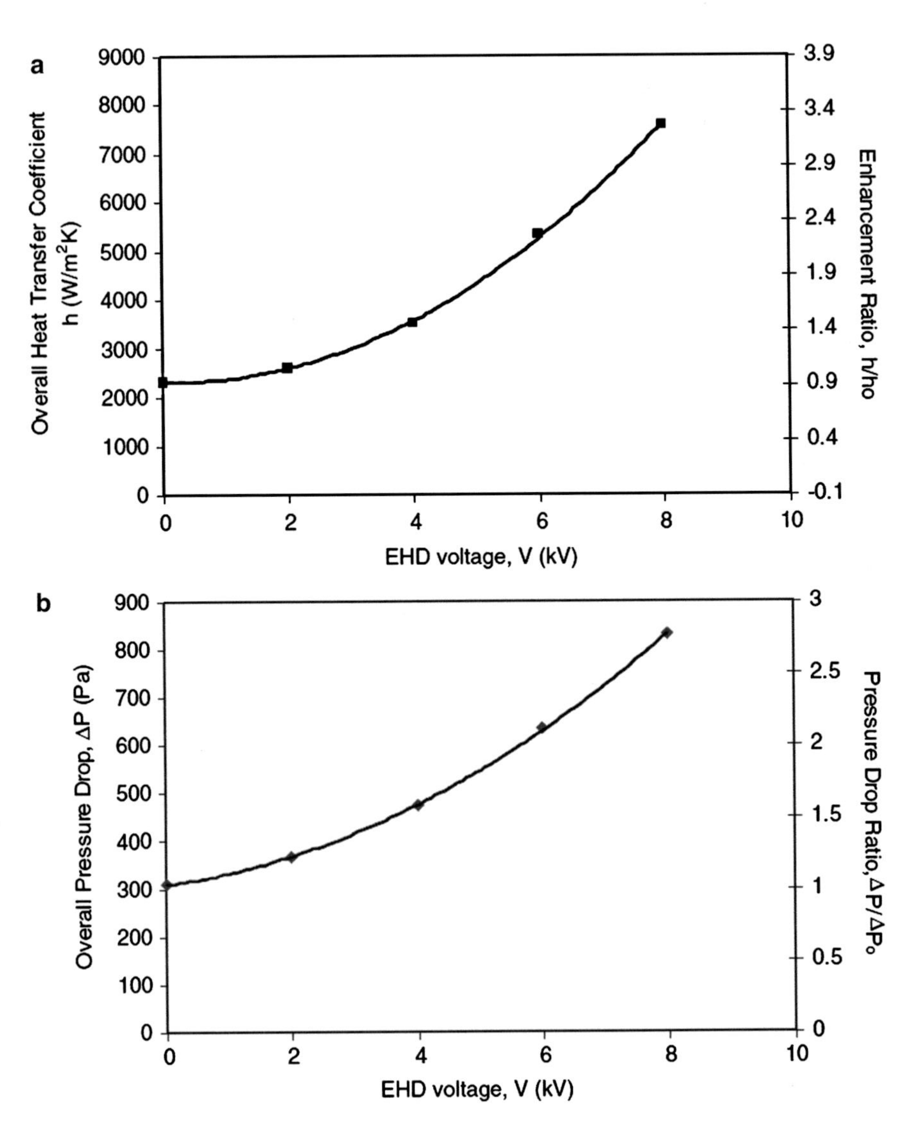

Fig. 10.1 Effects of applied voltage on heat transfer coefficient and pressure drop at mass flux of 83.4 kg/(m^2s), heat flux of 10.2 kW/m^2 and inlet quality of 66 %. (**a**) overall heat transfer coefficient, (**b**) overall pressure drop. Sadek et al. (2006)

enhanced by a factor up to 3.2 times for applied voltage of 8 kV. The pressure drop was increased by a factor of 1.5 at the same conditions of the maximum heat transfer enhancement. The EHD force extracts sufficient liquid from the liquid stratum at the bottom region to cause flow regime transition from stratified flow to annular flow. The decrease in the liquid layer thickness and introduction of droplets into the vapor core (Cotton et al. 2001) result in a large improvement in heat transfer coefficient and a relatively moderate increase in pressure drop.

Using the analogy to free convective flows, Cotton et al. (2001) and Cotton et al. (2005) proposed that the combined effects of electric and forced convection must be considered when $(M_d/\mathrm{Re_l}^2) \approx 1$, where M_d is the Masuda number or dielectric Rayleigh number and $\mathrm{Re_l}$ is the liquid Reynolds number. As shown in Fig. 10.2, if the inequalities $(M_d/\mathrm{Re_l}^2) \ll 1$ are satisfied, electric convection effects may be neglected, and conversely, if $(M_d/\mathrm{Re_l}^2) \gg 1$, forced convection effects may be neglected. This is exactly analogous to buoyancy driven flow and a similar argument might be made by comparing the Masuda number to the Grashof number in the absence of forced convection. This order of magnitude analysis is useful in determining the range and extent to which EHD may influence the flow and must be identified to determine the voltage levels required to induce liquid migration in order to affect heat transfer. The convective boiling flow pattern map was modified by the applied voltage due to liquid redistribution. The proposed flow pattern reconstruction with and without applied voltage by Cotton et al. (2005) is shown in Fig. 10.3. There is significant droplet entrainment and intermittent wetting due to climbing of liquid around the tube circumference as voltage increases. The increased interfacial electric force enhances the momentum suction pressure effect which attracts the liquid towards the upper part of the channel. With decreasing liquid layer thickness, interfacial waves are expected to increase due to the increase in

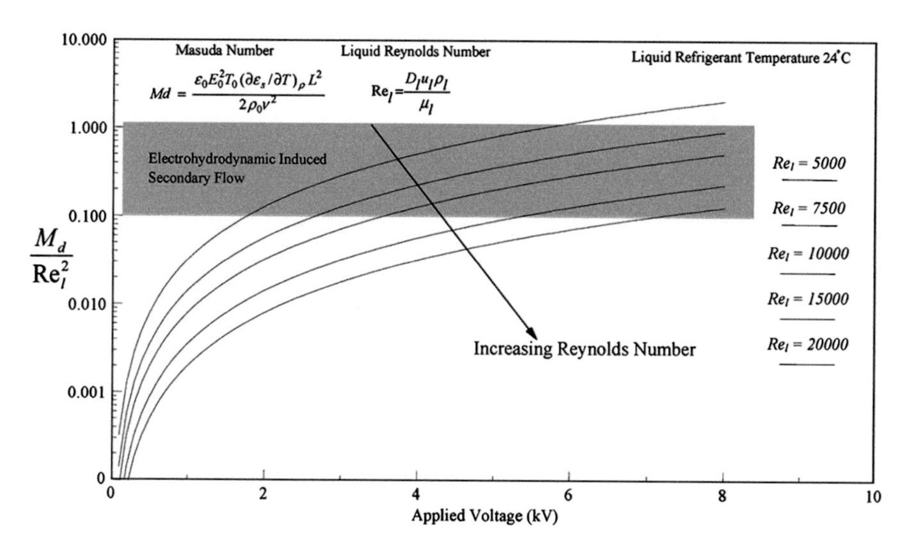

Fig. 10.2 The relationship of the Masuda number and Reynolds number ratio as a function of applied voltage. Cotton et al. (2005)

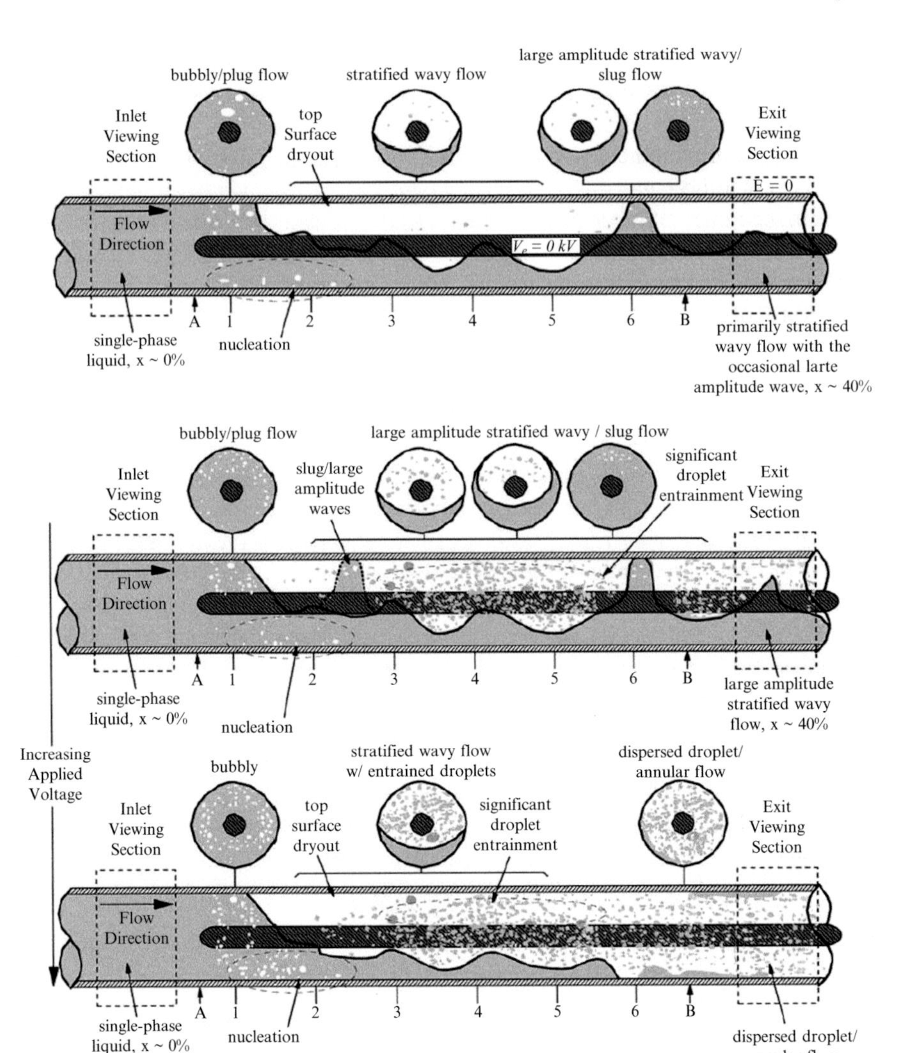

Fig. 10.3 Proposed reconstructed flow patterns for increasing DC voltage levels at a mass flux of 100 kg/(m^{2}s), a heat flux of 10 kW/m^2 and zero inlet quality. Cotton et al. (2005)

vapour velocity and increased nonuniformity of the electric field as more liquid is drawn upwards and entrained into the vapour core. By entraining liquid from the stratified layer upwards against gravity, through the interfacial electric forces, the thinner liquid layer increases the heat transfer coefficient along the bottom surface which in turn decreases the wall superheat. Further, redistributing the liquid to wet the upper portions of the channel that were previously dry, increases the heat transfer coefficient along the top surface which accounts for the decreases in the wall superheat.

Cheung et al. (1999) experimentally studied the EHD-assisted external condensation of R134a over a smooth tube under horizontal and vertical orientations. Their experimental results suggest a remarkable potential in utilizing EHD to enhance external condensation heat transfer. Figure 10.4 shows the effect of applied voltage on the average heat transfer coefficient at different electrode gaps (i.e., 0.8 mm, 1.6 mm, 3.2 mm). It is clear that the heat transfer coefficient increases with appiled voltage. The heat transfer enhancement is associated with strong EHD body force to remove the condensate from the heat transfer surface, thus reducing the overall thermal resistance and improving the heat transfer. Up to 620 % increase in heat transfer coefficient was achieved at an applied voltage of 26 kV. Besides, only a negligible amount of EHD power was required (only 0.06 % of the heat transfer rate) to achieve this large enhancement in heat transfer.

An effective cooling method for high temperature surfaces is to use liquid jets impingement on the heated surface (Zhou and Ma 2004; Cardenas and Narayanan 2012). As liquid jets approach the superheated surface, bubble will nucleate on the surface. A layer of alternately growing and collapsing bubbles in the immediate vicinity of the wall is acting as a source of turbulent mixing. The high degree of subcooling of the surrounding liquid and the impinging flow prevent the bubbles from detaching from the wall, limiting the entire boiling and condensation process to a boundary layer phenomenon. It can be assumed that a single bubble, while growing, will replace a hot fluid volume near the wall that is subsequently shifted into colder regions of the flow. Momentum and heat exchange of such a deplaced fluid volume with the main flow will result in a temperature drop in the fluid layer surrounding the bubble, leading to its collapse immediately after it has reached its maximum radius. Cold fluid will then be transported back to the heating wall and the process repeats. This process is clearly shown in Fig. 10.5.

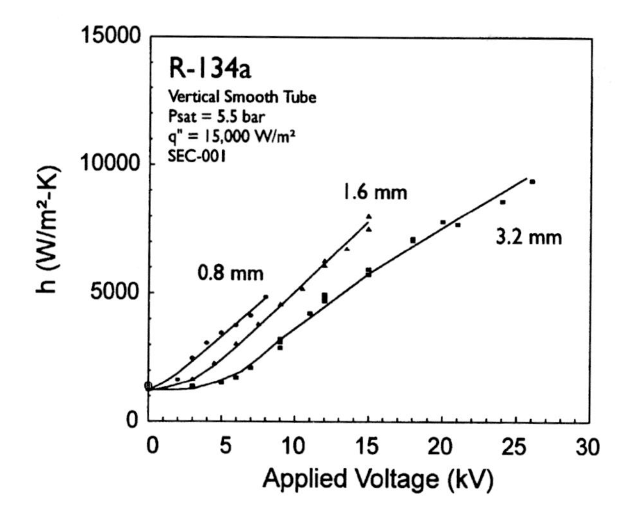

Fig. 10.4 Effect of applied voltage on heat transfer performance for vertical smooth tube external condensation. Cheung et al. (1999)

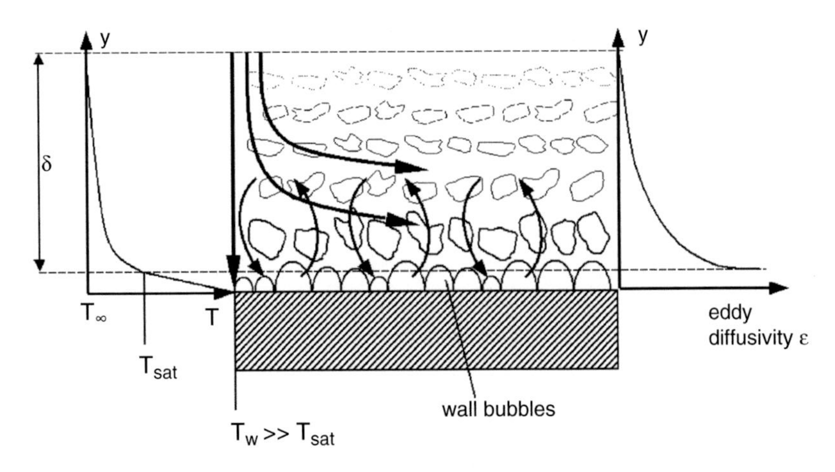

Fig. 10.5 Turbulence generation at the heating wall. Timm et al. (2003)

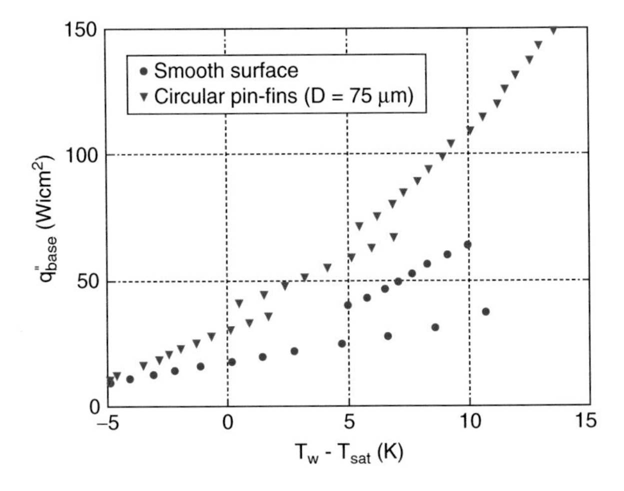

Fig. 10.6 Boiling curves for smooth and enhanced surfaces at a velocity of 2.2 m/s and an inlet temperature of 26 °C. Ndao et al. (2012)

Enhancement in the heat transfer coefficients for flow boiling jet impingement is generally attributed to the associated latent heat of vaporization and flow mixing induced by the violent departure on bubbles from the heated surface. To further enhance the performance of jet impingement, the introduction of highly enhanced structures on the impingement surface has been proposed (Lee and Vafai 1999; Ekkad and Kontrovitz 2002; Kim et al. 2009; Ndao et al. 2012). Flow boiling jet impingement on smooth and enhanced surfaces (circular micro pin fins, hydrofoil micro pin fins, and square micro pin fins) using R134a was carried out by Ndao et al. (2012). Jet impingement is characterized by the suppression of temperature overshoots. As shown in Fig. 10.6, transition from single-phase to the two-phase

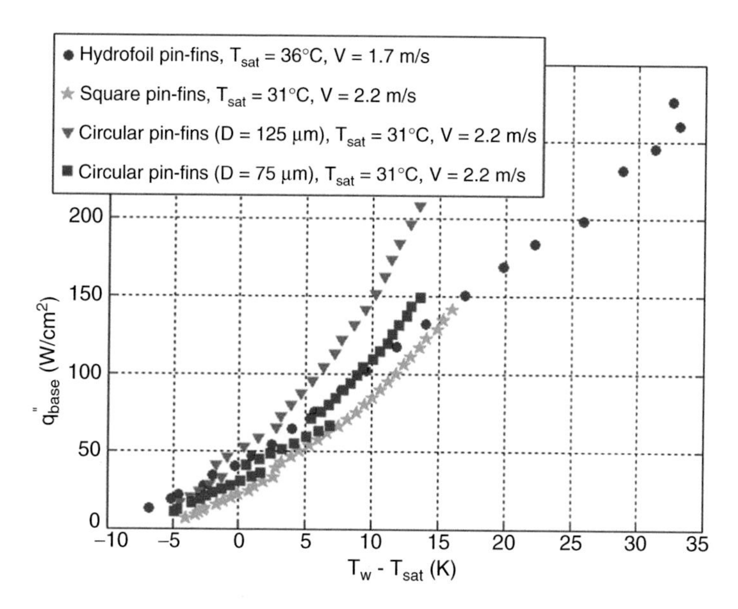

Fig. 10.7 Boiling curve based on the base area—effects of pin fin geometry. Ndao et al. (2012)

boiling occurred relatively smoothly, hence, suppressing any boiling hysteresis. When compared to smooth surfaces, boiling jet impingement on the micro pin fins accommodated higher heat fluxes for a given wall superheat. Two-phase heat transfer coefficients exceeding 150,000 W/(m^{2}K) were observed at a relatively low velocity of 2.2 m/s with the large circular micro pin fins (D = 125 μm) using R134a. Besides the suppression of temperature overshoots and higher heat transfer coefficients, the ONB was shown to occur at a relatively lower wall superheat when compared to flow boiling jet impingement on smooth surfaces. Figure 10.7 shows a comparison of the various micro pin fins considered in Ndao et al. (2012). The circular pin fins and the square pin fins showed better thermal performance than the hydrofoil pin fins. Note that the boiling curve of the hydrofoil pin fins appears to be higher than that of the square pin fins, which was due to the higher subcooling used in the former. Heat transfer enhancement was largely due to area enhancement and nucleate boiling enhancement with subcooled nucleate boiling being the dominant heat transfer mechanism.

Chapter 11
Additional Remarks

Flow Direction Flow direction directly affects the two-phase flow patterns, while flow pattern is closely related to heat transfer and pressure drop. Stratified flow is common in horizontal and inclined conventional channels under earth gravity conditions but not in vertical channels. Stratified flow is not an efficient heat transfer flow pattern. The thick liquid film in the bottom part of the horizontal channel presents a large thermal resistance, while for flow boiling, intermittent dryout might occur in the upper part of the channel. Various techniques can be used to redistribute the liquid film in stratified flow and thus enhance the heat transfer. For example, the EHD force, when appropriately applied on the two-phase flow, can carry liquid at the bottom part to wet the upper part of the heated surface. Microfin and herringbone tubes, can also push part of the liquid at the bottom surface to the upper and side surfaces, leading to earlier stratified-intermittent and intermittent-annular flow transitions, with increment in heat transfer.

Microgravity Phase change heat transfer is an attractive heat transfer mode for space applications under microgravity conditions. For boiling, buoyancy is no longer a bubble detachment force under reduced gravity, while surface tension becomes dominant in controlling bubble dynamics. On one hand, as the growing bubbles attached on the surface coalesce into primary large bubbles in microgravity during the boiling process, local dryout occurs easily with a low CHF level and inefficient heat removal. On the other hand, the dominant surface tension force alters the capillary effects, surface wetting and Marangoni effect in microgravity, which thus influence heat transfer and critical heat flux mechanisms (Zhang et al. 2014). For condensation, compared to earth gravity, the capillary length is much higher under microgravity conditions, and thus the droplet departure size is also much higher, which is very inefficient for heat transfer. A promising method to enhance phase change heat transfer is to use micro/nanostructures to modify the surface to alter the bubble/droplet dynamics under microgravity conditions. However, due to limited availability of microgravity opportunities and relatively strict experimental constraints, experimental activities on boiling in microgravity are still quite fragmentary,

S.K. Saha et al., *Advances in Heat Transfer Enhancement*,
DOI 10.1007/978-3-319-29480-3_11

not to mention heat transfer and CHF enhancement by micro/nanostructures in microgravity.

Flow Maldistribution Flow maldistribution causes deterioration of heat transfer (Rao Bobbili and Sundén 2008). The situation becomes more complicated for two-phase flows during boiling and condensation where uneven fluid distribution to the parallel channels or multi-channels induces pressure drop oscillations and reduces heat transfer and CHF. Especially in small channels with a hydraulic diameter comparable with the capillary length, the intensity of flow boiling instabilities is higher due to the higher rate of volumetric generation of vapor which induces considerable pressure drops. Rao Bobbili et al. (2006) evaluated the thermal performance of falling film plate condensers with flow maldistribution from port to channel considering the heat transfer coefficient inside the channels as a function of channel flow rate. As shown in Fig. 11.1, the thermal performance of the plate condenser (effectiveness) decreases as the flow maldistribution parameter m^2 increases. The value of m^2 approaches zero when the flow is uniformly distributed among the channels. The more flow maldistribution, the higher is the value of m^2. Therefore, a uniform flow distribution is important for heat transfer enhancement in two-phase flow. The flow unevenness can be mitigated by proper header designs and channel configurations.

Microchannel There are several distinct advantages for microchannels over conventional channels. Firstly, compactness and high heat-flux dissipation are required as the scale of the devices becomes small while the power density becomes large. Secondly, in the scaling-down from macro to microscale, the volume decreases with the third power of the characteristic linear dimensions, while surface area only

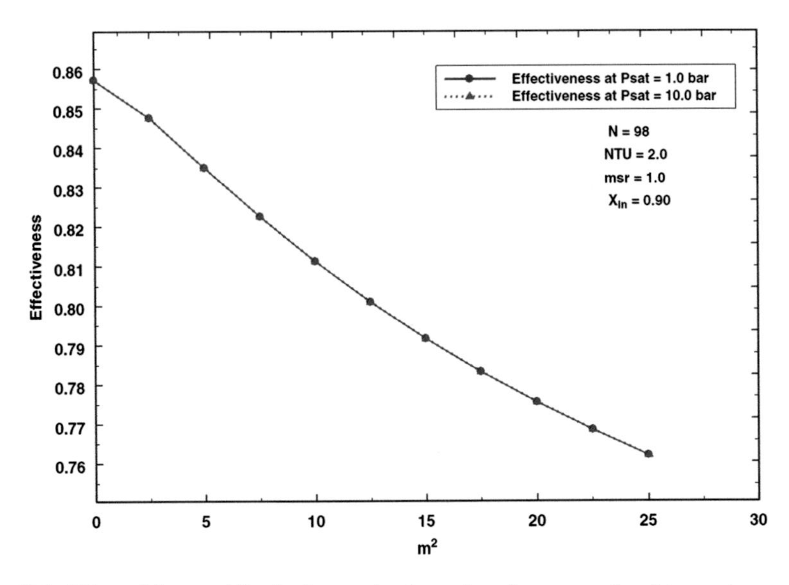

Fig. 11.1 Effect of flow maldistribution on the thermal performance of a plate condenser. Rao Bobbili et al. (2006)

decreases with the second power. In addition, fast fluid acceleration and close proximity of the bulk fluid to the wall surface in microchannels give high heat transfer coefficient values. However, microchannels also have several disadvantages when they are applied for two-phase flow. The pressure drop is comparably high (Li and Wu 2011). Besides, flow boiling instabilities induced by rapidly expanding confined and elongated bubbles in microchannels should be addressed carefully. Flow boiling instabilities can seriously modify the flow hydrodynamics, introduce transient surface temperature surges, hinder the thermal performance, generate vibrations, compromise structure integrity, and lead to premature initiation of the CHF. Two-phase flow in microchannels has been investigated almost 20 years. Interested readers may refer to reviews of Cheng et al. (2007), Ghiaasiaan (2008), Yarin et al. (2009), Kandlikar (2012) and Szczukiewicz et al. (2014) for further information. Further enhancement techniques and their characteristics for flow boiling heat transfer in microchannels can be found in Wu and Sundén (2014).

Fouling Fouling was and is still a major unresolved problem in heat transfer. Tackling fouling problems has gone hand-in-hand with improved heat transfer. Fouling restricts the heat transfer between fluids due to an additional fouling resistance and in general increases the pressure drop due to the rough surface formed by fouling. Fouling control can improve energy management and hence reduce the detrimental effects on heat transfer technology and the environment. Some of the enhancement techniques tend to prevent fouling, such as microfin tubes. However, the enhancement performance of some other enhancement techniques might be largely degraded by fouling. For example, fouling particles might pollute and partially fill the pores of porous coatings. Surface wettability of nanoscale coatings might be altered by fouling. These changes by fouling are probably unfavorable for phase-change heat transfer.

Chapter 12
Conclusions and Future Work

Increase in heat duty and demanding requirements on clean and sustainable technology are constantly moving the developments of heat transfer enhancement technology. This chapter presents an up-to-date overview of heat transfer enhancement techniques for two-phase flow (e.g., boiling and condensation) and mainly emphasizes those either commercially used enhancement techniques or the most recent enhancement methods. A special focus is on the enhancement technologies with a relatively low pressure drop penalty. Passive enhancement techniques such as surface coating, roughened and finned surfaces, insert devices, curved geometries and additives are highlighted. Several recent enhancement techniques, e.g., nanoscale surface coatings, microfin tubes and nanoparticle additives, are outlined for their promising potential in enhancing phase-change heat transfer, especially nanoscale surface coatings. Microchannels which can be considered as one of the promising passive enhancement techniques, are not detailed in this chapter as many aspects of microchannels have been covered in the recent literature. Among active enhancement techniques, electrohydrodynamic (EHD) phenomenon and jet impingement are briefly described.

As boiling and condensation are complex phenomena, heat transfer enhancement in two-phase flow calls for continued research and development. Several further remarks are outlined as follows.

- Nanoscale surface coatings show promising heat transfer enhancement and low pressure drop penalty. However, this enhancement technology is far from mature. There is still considerable controversy over what types of surface coatings can enhance both the heat transfer coefficient and CHF, not to mention the optimization of surface parameters. Although nanotechnology has greatly advanced, fabrication of nanoscale surface coatings with proper surface parameters, such as wettability and porosity, is still a big challenge. In addition, many nanostructures show aging effects, especially under flow conditions. Much work is still required on providing stable and durable nanoscale surface coatings for phase-change heat transfer enhancement.

S.K. Saha et al., *Advances in Heat Transfer Enhancement*,
DOI 10.1007/978-3-319-29480-3_12

- Hierarchical surface structures consisting of multiscale structures such as microstructures and nanostructures, if properly designed, might be a promising heat transfer enhancement method. More theoretical and experimental efforts are needed as phase-change phenomena on micro/nanostructures are very complicated due to the multiscales of the micro/nanostructures and the length-scale of the bubble/droplet dynamics.
- Although corrugated and finned surfaces are routinely used in industry, their enhancement performance can be further improved if special care is paid to fin/corrugation tip radius, which affects the surface tension and the condensate drainage. Also, future work is needed to optimize corrugation/fin parameters and configurations.
- The pressure drop penalty is relatively high for insert devices. Insert devices that can augment heat transfer and CHF with a moderate increase in pressure drop need to be developed.
- Nanoparticle addition has a complex effect on heat transfer performance. Future work may focus on obtaining direct information about bubble parameters, such as bubble departure diameter and frequency, during nanofluid flow boiling. These data may provide valuable insight in the mechanisms by which the nanoparticles affect the heat transfer coefficient.
- Development of new enhancement techniques can be based on and inspired by progresses in nanotechnology and bioengineering. Compound techniques might be investigated in the future.
- Many heat transfer enhancement techniques are still in the experimental stage. Additional work is recommended for commercialization efforts that will focus on technology transfers from the laboratory to the market place.
- As different flow patterns present different heat transfer mechanisms, enhancement techniques should be selected and optimized based on different flow patterns.
- Regarding to the application of heat transfer enhancement techniques, it raises concerns about their performance under fouling conditions. It will be interesting to develop enhancement techniques with anti-fouling features.

Acknowledgement Financial support from the Swedish National Research Council and the Swedish Energy Agency is gratefully acknowledged.

Bibliography

Akira Y, Hiroshi M (1988) Augmentation of convective and boiling heat transfer by applying an electro-hydrodynamical liquid jet. Int J Heat Mass Transf 31(2):407–417

Aly NH, Bedrose SD (1995) Enhanced film condensation of steam on spirally fluted tubes. Desalination 101(3):295–301

Ammerman CN, You SM (2001) Enhancing small-channel convective boiling performance using a microporous surface coating. ASME J Heat Transf 123(5):976–983

Attinger D, Frankiewicz C, Betz AR et al (2014) Surface engineering for phase change heat transfer: a review. MRS Energy Sustain A Rev J 1:E4

Bergles AE (1997) Heat transfer enhancement – the encouragement and accommodation of high heat fluxes. ASME J Heat Transf 119(1):8–19

Bergles AE (2002) ExHFT for fourth generation heat transfer technology. Exp Therm Fluid Sci 26(2):335–344

Bi J, Vafai K, Christopher DM (2015) Heat transfer characteristics and CHF prediction in nanofluid boiling. Int J Heat Mass Transf 80:256–265

Bocquet L, Lauga E (2011) A smooth future? Nat Mater 10(5):334–337

Boreyko JB, Chen CH (2009) Self-propelled dropwise condensate on superhydrophobic surfaces. Phys Rev Lett 103(18):184501

Bryan JE, Seyed-Yagoobi J (2001) Influence of flow regime, heat flux, and mass flux on electrohydrodynamically enhanced convective boiling. ASME J Heat Transf 123(2):355–367

Cardenas R, Narayanan V (2012) Heat transfer characteristics of submerged jet impingement boiling of saturated FC-72. Int J Heat Mass Transf 55(15):4217–4231

Celata GP, Cumo M, Mariani A (1994) Enhancement of CHF water subcooled flow boiling in tubes using helically coiled wires. Int J Heat Mass Transf 37(1):53–67

Chang JY, You SM (1997) Boiling heat transfer phenomena from microporous and porous surfaces in saturated FC-72. Int J Heat Mass Transf 40(18):4437–4447

Chen X, Wu J, Ma R et al (2011) Nanograssed micropyramidal architectures for continuous dropwise condensation. Adv Funct Mater 21(24):4617–4623

Cheng P, Wu HY, Hong FJ (2007) Phase-change heat transfer in microsystems. ASME J Heat Transf 129(2):101–108

Cheng J, Vandadi A, Chen CL (2012) Condensation heat transfer on two-tier superhydrophobic surfaces. Appl Phys Lett 101(13):131909

Cheung K, Ohadi MM, Dessiatoun SV (1999) EHD-assisted external condensation of R-134a on smooth horizontal and vertical tubes. Int J Heat Mass Transf 42(10):1747–1755

Cotton J, Shoukri M, Chang JS (2001) Oscillatory entrained droplet EHD two-phase flow. ASME J Heat Transf 123(4):622–622

Cotton J, Robinson AJ, Shoukri M et al (2005) A two-phase flow pattern map for annular channels under a DC applied voltage and the application to electrohydrodynamic convective boiling analysis. Int J Heat Mass Transf 48(25):5563–5579

Daniel S, Chaudhury MK, Chen JC (2001) Fast drop movements resulting from the phase change on a gradient surface. Science 291(5504):633–636

Dawidowicz B, Cieśliński JT (2012) Heat transfer and pressure drop during flow boiling of pure refrigerants and refrigerant/oil mixtures in tube with porous coating. Int J Heat Mass Transf 55(9):2549–2558

Djordjevic E, Kabelac S (2008) Flow boiling of R134a and ammonia in a plate heat exchanger. Int J Heat Mass Transf 51(25):6235–6242

Ekkad SV, Kontrovitz D (2002) Jet impingement heat transfer on dimpled target surfaces. Int J Heat Fluid Flow 23:22–28

Feng Z, Wu Z, Li W, Sundén B (2016) Effect of surfactant on flow boiling heat transfer of ethylene glycol/water mixtures in a mini-tube. Heat Transfer Eng. doi:10.1080/01457632.2015.1111112

Furberg R, Palm B, Li S et al (2009) The use of a nano-and microporous surface layer to enhance boiling in a plate heat exchanger. ASME J Heat Transf 131:101010

Garimella S, Richards DE, Christensen RN (1988) Experimental investigation of heat transfer in coiled annular ducts. ASME J Heat Transf 110(2):329–336

Ghiaasiaan SM (2008) Two-phase flow: boiling and condensation in convective and miniature systems. Cambridge University Press, New York

Gregorig R (1954) Hautcondensation an feingewelten Oberflachen bei Beruksichtigung der Oberflachenspannungen. Z Angew Math Phys 5:36–49

Guo SP, Wu Z, Li W et al (2015) Condensation and evaporation heat transfer characteristics in horizontal smooth, herringbone and enhanced surface EHT tubes. Int J Heat Mass Transf 85:281–291

Hanlon MA, Ma HB (2003) Evaporation heat transfer in sintered porous media. ASME J Heat Transf 125(4):644–652

Hata K, Masuzaki S (2011) Heat transfer and critical heat flux of subcooled water flow boiling in a SUS304-tube with twisted-tape insert. ASME J Therm Sci Eng Appl 3(1):012001

Henderson K, Park YG, Liu L et al (2010) Flow-boiling heat transfer of R-134a-based nanofluids in a horizontal tube. Int J Heat Mass Transf 53(5):944–951

Hsieh YY, Lin TF (2002) Saturated flow boiling heat transfer and pressure drop of refrigerant R-410A in a vertical plate heat exchanger. Int J Heat Mass Transf 45(5):1033–1044

Hsieh SS, Jang KJ, Tsai HH (2003) Evaporative characteristics of R-134a and R-600a in horizontal tubes with perforated strip-type inserts. Int J Heat Mass Transf 46(10):1861–1872

Hsu YY (1962) On the size range of active nucleation cavities on a heating surface. ASME J Heat Transf 84(3):207–213

Kandlikar SG (2010) Scale effects on flow boiling heat transfer in microchannels: a fundamental perspective. Int J Therm Sci 49:1073–1085

Kandlikar SG (2012) History, advances, and challenges in liquid flow and flow boiling heat transfer in microchannels: a critical review. ASME J Heat Transf 134(3):034001

Kang SH, Wu N, Grinthal A et al (2011) Meniscus lithography: evaporation-induced self-organization of pillar arrays into Moiré patterns. Phys Rev Lett 107(17):177802

Kedzierski MA, Goncalves JM (1999) Horizontal convective condensation of alternative refrigerants within a micro-fin tube. J Enhanc Heat Transf 6(2–4):161–178

Khanikar V, Mudawar I, Fisher T (2009) Effects of carbon nanotube coating on flow boiling in a micro-channel. Int J Heat Mass Transf 52(15):3805–3817

Kim KJ, Lefsaker AM, Razani A et al (2001) The effective use of heat transfer additives for steam condensation. Appl Therm Eng 21(18):1863–1874

Kim SJ, McKrell T, Buongiorno J et al (2008) Alumina nanoparticles enhance the flow boiling critical heat flux of water at low pressure. ASME J Heat Transf 130(4):044501

Kim DK, Kim SJ, Bae JK (2009) Comparison of thermal performances of plate-fin and pin-fin heat sinks subject to an impinging flow. Int J Heat Mass Transf 52:3510–3517

Kim H, Ahn HS, Kim MH (2010a) On the mechanism of pool boiling critical heat flux enhancement in nanofluids. ASME J Heat Transf 132(6):061501

Kim SJ, McKrell T, Buongiorno J et al (2010b) Subcooled flow boiling heat transfer of dilute alumina, zinc oxide, and diamond nanofluids at atmospheric pressure. Nucl Eng Des 240(5):1186–1194

Kousalya AS, Hunter CN, Putnam SA et al (2012) Photonically enhanced flow boiling in a channel coated with carbon nanotubes. Appl Phys Lett 100(7):071601

Krishna KH, Ganapathy H, Sateesh G et al (2011) Pool boiling characteristics of metallic nanofluids. ASME J Heat Transf 133:111501

Kumar CS, Suresh S, Yang L et al (2014) Flow boiling heat transfer enhancement using carbon nanotube coatings. Appl Therm Eng 65(1):166–175

Kumar CS, Suresh S, Praveen AS et al (2016) Effect of surfactant addition on hydrophilicity of ZnO-Al_2O_3 composite and enhancement of flow boiling heat transfer. Exp Therm Fluid Sci 70:325–334

Kuo WS, Lie YM, Hsieh YY et al (2005) Condensation heat transfer and pressure drop of refrigerant R-410A flow in a vertical plate heat exchanger. Int J Heat Mass Transf 48(25):5205–5220

Lee J, Mudawar I (2007) Assessment of the effectiveness of nanofluids for single-phase and two-phase heat transfer in micro-channels. Int J Heat Mass Transf 50(3):452–463

Lee DY, Vafai K (1999) Comparative analysis of jet impingement and microchannel cooling for high heat flux applications. Int J Heat Mass Transf 42:1555–1568

Lee H, Li S, Hwang Y et al (2013) Experimental investigations on flow boiling heat transfer in plate heat exchanger at low mass flux condition. Appl Therm Eng 61(2):408–415

Li C, Peterson GP (2007) Parametric study of pool boiling on horizontal highly conductive microporous coated surfaces. ASME J Heat Transf 129:1465–1475

Li W, Wu Z (2011) Generalized adiabatic pressure drop correlations in evaporative micro/minichannels. Exp Therm Fluid Sci 35(6):866–872

Li GQ, Wu Z, Li W et al (2012) Experimental investigation of condensation in micro-fin tubes of different geometries. Exp Therm Fluid Sci 37:19–28

Liebenberg L, Meyer JP (2007) In-tube passive heat transfer enhancement in the process industry. Appl Therm Eng 27(16):2713–2726

Liter SG, Kaviany M (2001) Pool-boiling CHF enhancement by modulated porous-layer coating: theory and experiment. Int J Heat Mass Transf 44(22):4287–4311

Liu D, Yu L (2011) Single-phase thermal transport of nanofluids in a minichannel. ASME J Heat Transf 133(3):031009

Liu TQ, Sun W, Sun XY et al (2012) Mechanism study of condensed drops jumping on super-hydrophobic surfaces. Colloids Surf A Physicochem Eng Asp 414:366–374

Longo GA, Zilio C, Righetti G et al (2014) Condensation of the low GWP refrigerant HFO1234ze (E) inside a Brazed Plate Heat Exchanger. Int J Refrig 38:250–259

Lopina RF, Bergles AE (1973) Subcooled boiling of water in tape-generated swirl flow. ASME J Heat Transf 95(2):281–283

Ma X, Briggs A, Rose JW (2004) Heat transfer and pressure drop characteristics for condensation of R113 in a vertical micro-finned tube with wire insert. Int Commun Heat Mass Transfer 31(5):619–627

Ma A, Wei J, Yuan M et al (2009) Enhanced flow boiling heat transfer of FC-72 on micro-pin-finned surfaces. Int J Heat Mass Transf 52(13):2925–2931

Manglik RM, Bergles AE (2013) Characterization of twisted-tape-induced helical swirl flows for enhancement of forced convective heat transfer in single-phase and two-phase flows. ASME J Therm Sci Eng Appl 5(2):021010

Marto PJ (1986) Recent progress in enhancing film condensation heat transfer on horizontal tubes. Heat Transfer Eng 7(3–4):53–63

Melcher CL (1981) Thermoluminescence of meteorites and their terrestrial ages. Geochim Cosmochim Acta 45(5):615–626

Miljkovic N, Wang EN (2013) Condensation heat transfer on superhydrophobic surfaces. MRS Bull 38(5):397–406

Miljkovic N, Enright R, Wang EN (2013) Modeling and optimization of superhydrophobic condensation. ASME J Heat Transf 135(11):111004

Mozafari M, Akhavan-Behabadi MA, Qobadi-Arfaee H et al (2015) Experimental study on condensation flow patterns inside inclined U-bend tubes. Exp Therm Fluid Sci 68:276–287

Narhe RD, Khandkar MD, Shelke PB et al (2009) Condensation-induced jumping water drops. Phys Rev E 80(3):031604

Ndao S, Peles Y, Jensen MK (2012) Experimental investigation of flow boiling heat transfer of jet impingement on smooth and micro structured surfaces. Int J Heat Mass Transf 55(19):5093–5101

Nnanna AG (2007) Experimental model of temperature-driven nanofluid. ASME J Heat Transf 129(6):697–704

Owhadi A, Bell KJ, Crain B (1968) Forced convection boiling inside helically-coiled tubes. Int J Heat Mass Transf 11(12):1779–1793

Patankar NA (2010) Supernucleating surfaces for nucleate boiling and dropwise condensation heat transfer. Soft Matter 6(8):1613–1620

Peng H, Ding G, Hu H (2011) Effect of surfactant additives on nucleate pool boiling heat transfer of refrigerant-based nanofluid. Exp Therm Fluid Sci 35(6):960–970

Qu ZG, Xu ZG, Zhao CY et al (2012) Experimental study of pool boiling heat transfer on horizontal metallic foam surface with crossing and single-directional V-shaped groove in saturated water. Int J Multiphase Flow 41:44–55

Rainey KN, Li G, You SM (2001) Flow boiling heat transfer from plain and microporous coated surfaces in subcooled FC-72. ASME J Heat Transf 123(5):918–925

Rao Bobbili PR, Sundén B (2008) Steam condensation in parallel channels of plate heat exchangers – an experimental investigation. Heat Transfer Res 39(3):197–210

Rao Bobbili PR, Sundén B, Das SK (2006) Thermal analysis of plate condensers in presence of flow maldistribution. Int J Heat Mass Transf 49(25):4966–4977

Reay D, Ramshaw C, Harvey A (2013) Process intensification: engineering for efficiency, sustainability and flexibility. Butterworth-Heinemann, Oxford

Rohsenow WM (1998) Handbook of heat transfer, vol 3. McGraw-Hill, New York
Rose JW (2004) Surface tension effects and enhancement of condensation heat transfer. Chem Eng Res Des 82(4):419–429
Rykaczewski K, Scott JHJ (2011) Methodology for imaging nano-to-microscale water condensation dynamics on complex nanostructures. ACS Nano 5:5962–5968
Sadek H, Robinson AJ, Cotton JS et al (2006) Electrohydrodynamic enhancement of in-tube convective condensation heat transfer. Int J Heat Mass Transf 49(9):1647–1657
Santini L, Cioncolini A, Butel MT et al (2016) Flow boiling heat transfer in a helically coiled steam generator for nuclear power applications. Int J Heat Mass Transf 92:91–99
Sarwar MS, Jeong YH, Chang SH (2007) Subcooled flow boiling CHF enhancement with porous surface coatings. Int J Heat Mass Transf 50(17):3649–3657
Seyed-Yagoobi J, Bryan JE (1999) Enhancement of heat transfer and mass transport in single-phase and two-phase flows with electrohydrodynamics. Adv Heat Tran 33:95–186
Singh N, Sathyamurthy V, Peterson W et al (2010) Flow boiling enhancement on a horizontal heater using carbon nanotube coatings. Int J Heat Fluid Flow 31(2):201–207
Sterner D, Sundén B (2006) Performance of plate heat exchangers for evaporation of ammonia. Heat Transfer Eng 27(5):45–55
Straub J (1994) The role of surface tension for two-phase heat and mass transfer in the absence of gravity. Exp Therm Fluid Sci 9(3):253–273
Sundén B (2012) Introduction to heat transfer. WIT Press, Southampton
Sundén B, Wu Z (2015) Advanced heat exchangers for clean and sustainable technology. In: Yan J (ed) Handbook of clean energy systems. Wiley, New York
Szczukiewicz S, Magnini M, Thome JR (2014) Proposed models, ongoing experiments, and latest numerical simulations of microchannel two-phase flow boiling. Int J Multiphase Flow 59:84–101
Thome JR (1990) Enhanced boiling heat transfer. Hemisphere, New York
Thome JR, Dupont V, Jacobi AM (2004) Heat transfer model for evaporation in microchannels. Part I: presentation of the model. Int J Heat Mass Transf 47(14):3375–3385
Timm W, Weinzierl K, Leipertz A (2003) Heat transfer in subcooled jet impingement boiling at high wall temperatures. Int J Heat Mass Transf 46(8):1385–1393
Ujereh S, Fisher T, Mudawar I (2007) Effects of carbon nanotube arrays on nucleate pool boiling. Int J Heat Mass Transf 50(19):4023–4038
Vafaei S, Wen D (2010) Critical heat flux (CHF) of subcooled flow boiling of alumina nanofluids in a horizontal microchannel. ASME J Heat Transf 132(10):102404
Wang CH, Dhir VK (1993) Effect of surface wettability on active nucleation site density during pool boiling of water on a vertical surface. ASME J Heat Transf 115:659–669
Wang LK, Sunden B, Yang QS (1999) Pressure drop analysis of steam condensation in a plate heat exchanger. Heat Transfer Eng 20(1):71–77
Wang H, Garimella SV, Murthy JY (2007a) Characteristics of an evaporating thin film in a microchannel. Int J Heat Mass Transf 50(19):3933–3942
Wang LK, Sunden B, Manglik RM (2007b) Plate heat exchangers: design, applications and performance. WIT Press, Southampton
Wanniarachchi AS, Marto PJ, Rose JW (1986) Film condensation of steam on horizontal finned tubes: effect of fin spacing. ASME J Heat Transf 108(4):960–966
Webb RL (1983) Nucleate boiling on porous coated surfaces. Heat Transfer Eng 4(3–4):71–82
Webb RL (2004) Donald Q. Kern Lecture Award Paper: Odyssey of the enhanced boiling surface. ASME J Heat Transf 126(6):1051–1059
Webb RL, Kim NH (2005) Principles of enhanced heat transfer. Taylor & Francis, New York
Wellsandt S, Vamling L (2005) Prediction method for flow boiling heat transfer in a herringbone microfin tube. Int J Refrig 28(6):912–920
Wörner M (2003) A compact introduction to the numerical modeling of multiphase flows. Forschungszentrum, Karlsruhe
Wu Z, Sundén B (2014) On further enhancement of single-phase and flow boiling heat transfer in micro/minichannels. Renew Sustain Energy Rev 40:11–27

Wu Z, Sundén B (2015a) Frictional pressure drop correlations for single-phase flow, condensation, and evaporation in microfin tubes. ASME J Heat Transf 138(2):022901

Wu Z, Sundén B (2015b) Flow-pattern based heat transfer correlations for stable flow boiling in micro/minichannels. ASME J Heat Transf 138:031501

Wu Z, Sundén B (2015c) Heat transfer correlations for elongated bubbly flow in flow boiling micro/minichannels. Heat Transfer Eng. doi:10.1080/01457632.2015.1098269

Wu Z, Sundén B, Li W et al (2013a) Evaporative annular flow in micro/minichannels: a simple heat transfer model. ASME J Therm Sci Eng Appl 5:031009

Wu Z, Wu Y, Sundén B et al (2013b) Convective vaporization in micro-fin tubes of different geometries. Exp Therm Fluid Sci 44:398–408

Wu Z, Sundén B, Wang L et al (2014) Convective condensation inside horizontal smooth and microfin tubes. ASME J Heat Transf 136(5):051504

Wu Z, Sundén B, Wadekar VV et al (2015) Heat transfer correlations for single-phase flow, condensation, and boiling in microfin tubes. Heat Transfer Eng 36(6):582–595

Xu L, Xu J (2012) Nanofluid stabilizes and enhances convective boiling heat transfer in a single microchannel. Int J Heat Mass Transf 55(21):5673–5686

Yang XF, Liu ZH (2012) Flow boiling heat transfer in the evaporator of a loop thermosyphon operating with CuO based aqueous nanofluid. Int J Heat Mass Transf 55(25):7375–7384

Yang CY, Webb RL (1997) A predictive model for condensation in small hydraulic diameter tubes having axial micro-fins. ASME J Heat Transf 119(4):776–782

Yarin LP, Mosyak A, Hetsroni G (2009) Fluid flow. heat transfer and boiling in micro-channels. Springer, Berlin

Yu L, Sur A, Liu D (2015) Flow boiling heat transfer and two-phase flow instability of nanofluids in a minichannel. ASME J Heat Transf 137(5):051502

Zhang Y, Wei J, Xue Y et al (2014) Bubble dynamics in nucleate pool boiling on micro-pin-finned surfaces in microgravity. Appl Therm Eng 70(1):172–182

Zhao L, Guo L, Bai B et al (2003) Convective boiling heat transfer and two-phase flow characteristics inside a small horizontal helically coiled tubing once-through steam generator. Int J Heat Mass Transf 46(25):4779–4788

Zhou DW, Ma CF (2004) Local jet impingement boiling heat transfer with R113. Heat Mass Transf 40(6–7):539–549